# FAIRGROUND ENGINES IN FOCUS

by John Crawley

PUBLISHED BY
JOHN CRAWLEY LTD, TURVEY, BEDFORDSHIRE

First published in 1983 by
John Crawley Limited
Field House, Turvey
Bedfordshire MK43 8DU

Printed in Great Britain by
AB Printers Limited, Leicester



ISBN 0 9508046 1 4

Reprinted 1984

To the late Stanley Thurston
and his unknown driver.

# FOREWORD

by W.T. Woolrych, A.M.B.I.M.
General Secretary of
The Showman's Guild of Great Britain.

As you pass the showman of today on the road, towing his loads behind purpose-built diesel locomotives or massive six- and eight-wheeled lorries, it is hard to realise that this is a comparatively novel sight.

Although showmen were quick to exploit the potential of the mass of ex-W.D. petrol-engined lorries which became available after the first world war, the heavier loads were drawn by steam engines for many more years and some remained in service until the 1950s. Indeed had the second world war not intervened, forcing these magnificent machines into idleness while simultaneously accelerating the development of compression ignition engines, it is possible that some would have graced the fairgrounds for longer still, to enchant and captivate another generation of seven-year olds like John Crawley.

In this meticulously researched book with its splendid photographs, Mr Crawley not only reminds us of the might and power of these once-familiar leviathans; he has also captured the atmosphere of the way of life they epitomised. But this way of life is far from dead, and it is as much for its spirit of continuity as with its sense of history that this book is to be commended.

Here are the great names of the past: Burrell, Savage, Fowler and Garrett; but here too are names which are no less great today. Names such as Thurston, Whitelegg, Collins, Anderton & Rowland, Heal, Evans, Holland, Taylor, Traylen and Manning to select but a few. These families still stand at the forefront of showland, and thus the present and the past are interwoven in a living tapestry of tradition, a tradition that is the sessence of the showman's calling.

Between these pages too, are glimpses of the day to day problems faced by showmen at the time, problems which have their exact counterparts today. Removing the caked and glutinous mud after winching off some saturated meadow; "heavy recovery" after a mishap on the road; finding somewhere to stop while waiting to pull on to a fairground; and manoeuvring the loads into tight positions at street fairs, without damaging the buildings or scratching the paintwork; all are faithfully recorded, and will be no less familiar to today's showmen than they were to their parents and grandparents.

The showman's ability as a driver is one of the most remarkable and in many ways the least acknowledged of his skills, and perhaps in part accounts for the comparatively large number of showmen's engines which are still with us today, lovingly preserved and proudly displayed by their enthusiastic new owners at steam fairs and rallies. For the showman is not only carpenter and decorator, electrician and mechanic; he is above all a superb driver of heavy equipment, equipment which he owns and with which he earns his living. And so that this equipment shall safely reach the next fair, to be built up and ready to open by the appointed time, the vehicles with which the showman tows or transports that equipment are lovingly cared for and meticuously maintained, and this was never more so than with the true showmans engine, with its full length canopy, 'barley sugar' brass and rich paintwork.

Another feature which struck me as I turned the pages is the astonishing length of the trains of wagons and trailers which were towed behind these engines. To be sure, they did not travel very fast, but John Crawley tells me of a photograph of one such train, sadly not reproduced in this book, where the loads consisted of thirteen such trailers and may be seen snaking their way around three streets in one continuous train. Such a sight would be impossible today, where the showman is bound like all other road users by the provisions of the Motor Vehicles (Construction & Use) Regulations, but it would be interesting to see what a contemporary traffic policeman would make of an E.R.F or A.E.C. towing a trailer train nearly fifty yards long! On the other hand, some legislation which remains with us today is clearly a direct legacy from the days of steam, legislation such as that contained in the Road Traffic Regulation Act 1967, which still restricts the maximum speed of a locomotive drawing two or more trailers to 12 miles per hour.

This then is the special appeal — to me at least — of this book; the link between past and present and the record of workaday activities in the showman's life. In consequence the book commends itself to showmen — many of whom will see their own parents and grandparents depicted in these pages — historians, students of steam engineering and the general reader alike.

W.T. Woolrych

26th April 1983.

# INTRODUCTION

The life of travelling showmen was hard, with long hours, dirty work, and little chance of staying in the same place for more than a few days at a time. Their engines were principally the means by which they earned their living.

New engines were delivered in the distinctive colours so beloved by their owners, and often with a great deal of ornamental brasswork, but the hard work and frequent abuse that they suffered soon brought about a deterioration in appearance. They were not regarded with affection by the fairground public, and, very differently from today, were more often than not tucked away out of sight behind the side stalls unless they had to be alongside the ride they were driving.

I grew up by a main road that saw much of the the travelling showmen: very near my parents' home was a stream where they would stop to replenish their engines.

On one such occasion, when I was about seven, I was invited by a driver to jump up for a quick look round the footplate of Stanley Thurston's Burrell 'Britannia', an innocent enough gesture that would spark off a lifelong interest leading to the ownership in later years of four such engines at various times, all of course in preservation.

The identity of the Burrell's driver is now lost in time, but my wife has said on more than one occasion that she would like a chance to meet him!

When reading the photo captions, the following points should be borne in mind: the month quoted as the date of building refers to the date at which the engine was available for delivery, construction having taken several months; when the term 'road locomotive' is used, it refers to a plain haulage engine as distinct from the showmans engine which differed in having a full-length awning, dynamo and very often, twisted brass columns, although technically speaking they were both road locomotives; when a showman is referred to as converting a machine to a showmans engine, it does not necessary mean that he did the work himself, although many did, but that he was responsible for having the conversion made; and the small black square ■ after an engine number indicates that it has survived and is now in preservation. It should also be remembered that the registration numbers carried by these machines were not necessarily allocated on purchase but might, for example, on an engine built around 1900, have been obtained as much as twenty years later when the Road Traffic Act came into force in 1921.

I would like to thank my many friends who lent photographs, which are all individually acknowledged. Copies of those loaned by the Leeds and District Traction Engine Club may be purchased by writing to the Club at 5 Knightscroft Avenue, Rothwell, Leeds LS26 0PF.

To Alan Duke go my sincere thanks for his patience in answering numerous questions of identity and ownership, and also to Reg Smith whose knowledge of showmans engines and the showland families is surely unsurpassed.

Finally I must thank my friends and the public for their reception of my previous book, which has given me the courage to go ahead and present *Fairground Engines in Focus.*

John Crawley

# LIST OF PLATES

1. Allchin 8nhp single cylinder engine No 1246 was built in May 1903 and sold to Fred Gray of Hampstead. She was used for travelling his electric galloping horses roundabout and later sold to William Twigg of Matlock. Pictured here with the galloper loads at Markyate, Hertfordshire, in 1904.
*(courtesy Leeds T.E. Club)*

2. A bridge at Allonby, near Maryport, Cumbria gives way beneath Aveling & Porter 8nhp compound engine No 5635, *His Majesty*, Registration No BL 1998. Built in February 1905, the year this photograph was taken, it was sold to John Caris & Sons of Heaton, Northumberland.
*(courtesy Leeds T.E. Club)*

3. Aveling & Porter compound tractor No 9431, Registration No NO 5950, was built in July 1922 and sold to Essex County Council. She was eventually sold to H. Gumble of Willingale, Essex with whom she spent the rest of her working life. The canopy sides were lettered 'British, John Gumble's, Amusements'.
*(author's collection)*

4. An Aveling & Porter traction engine owned by amusement caterer Sidney White caught in the floods at Merthyr Tydfil in 1907. No further details are known as in the early days showmen purchased second hand traction engines which were a great improvement over the horses upon which they had hitherto been dependent. Many of these were very old and had changed owners so many times details of the transactions have not always survived.

*(author's collection)*

5. Aveling & Porter 6nhp compound engine No 9419, Registration No PD 5504, was built in September 1922 as a road locomotive and sold to W.H. Glaze & Sons Ltd of Kingston. Sometime later in her life she was sold to Sam Jones of Chertsey, Surrey, who converted her into a showmans engine and travelled her for the rest of her days until she was sold to J.W. Hardwick & Sons of West Ewell for scrapping. Seen here standing derelict at Chertsey on Saturday, 13th May 1939.

*(courtesy A.C. Durrant)*

6. Burrell 8nhp single crank compound engine No 1628, Registration No CF 3634, built in August 1892 as a road locomotive and sold to E.P. Bailey of Cranbrook, Kent. In 1894 she was returned to Burrells works where she was converted into a showmans engine and sold to Joseph Matthews of Haslemere, Surrey. Early in 1904 she was sold to Charles Barnes of Great Ryburgh, Norfolk, who named her *Royal Hunter* and used her for travelling his set of galloping horses roundabout. It was in his ownership that this incident occured on the quay at Burnham Overy Regatta Fair in 1904. He finally sold her out of showland in 1918.

*(author's collection)*

7. Pictured at Redruth Fair in 1898 is Burrell 10nhp compound engine No 1740, *Cornishman*, Registration No CO 3885. Built in March 1894 and sold to W.C. & S. Hancock of Bristol, she spent her working life travelling the firm's Savage four-abreast set of galloping horses. In 1926 engine and ride were bought by Sam Manning of Enfield who scrapped both the ride and engine in 1934.
*(author's collection)*

8. This accident involving Burrell No 1740 *Cornishman'*, occurred at Weir Cross Hill, Newton Abbott in 1912. It appears that the engine ran away down the hill, crashed through the hedge and turned on her side with the loads following.
*(author's collection)*

9. Burrell 8nhp single crank compound engine No 1796, *Lady Combermere*, Registration No CF 3758. Built in September 1894 and sold to Charles Farrell of Chester, she was later acquired by A. Dobson of Chester but the date is not recorded. She travelled his set of galloping horses roundabout until 1909 when whe was sold out of showland.
*(courtesy Leeds T.E. Club)*

10. Burrell 8nhp single crank compound engine No 1810, *Clarence*, Registration No BE 7332, was built in February 1895 and sold to Charles Warren of Lincoln for travelling his Wild West Broncho pony ride. Later she was owned by W. Stock of Bilsthorpe, Notts and by 1921 was with John Kendal of Loughton, Lindsey until she was scrapped in December 1924. *(author's collection)*

11. Burrell 10nhp compound engine No 1908, *Excelsior*, was built in March 1896 and sold to Harry J. Wallis Junior of Seaforth, Lancashire. She was one of the first Burrells to be fitted with wooden blocks on the rear wheels as an aid to adhesion. *(courtesy B.D. Stoyel)*

12. Seen at Lindfield Fair, Sussex on 12th August 1939 is Burrell 10nhp compound engine No 2134 *Wanderer*, Registration No BW 6546. Built in October 1898 as a road locomotive she was sold to A. Wickens of Reading and had two more owners before she was bought by Tom Smith of Shoreham by Sea, West Sussex who converted her into a showmans engine and named her. She was scrapped in 1950. *(author's collection)*

13. Burrell 8nhp compound engine No 2268, *The Councillor*, Registration No WR 6913 was built in March 1900 and sold to G.T. Tuby & Sons of Doncaster. This was the first engine to be acquired by the firm which was to become well known in and around their native Yorkshire. She was painted in Great Eastern Railway blue with yellow wheels. For a number of years she worked with Tuby's galloping horses until, in the nineteen-thirties, she was sold to W. North & Sons of Denaby in the West Riding during which time she received a new Registration No ET 3402. Her last owner was John T. North of Parkgate also in the West Riding of Yorkshire. Pictured here at Kirkheaton Fair in June 2936 whilst in the ownership of John T. North. *(author's collection)*

14. Burrell 8nhp compound engine No 2281, Registration No HO 5910 was built in May 1900 as a road locomotive and sold to Smith Bros of Walsall, Staffordshire. In June 1908 she was sold to J.W. Chipperfield of Redditch, Worcestershire who converted her into a showmans engine and named her *Queen of the Midlands*. Photographed in 1909 at the Mop Fair, Southam, Gloucestershire. *(courtesy S. White)*

15. Burrell 8nhp compound No 2350, *Pride of Wales*, Registration No FK 1121 was built in February 1901 and sold to William Ford of Brynmawr, Brecon. In 1908 she was sold to F. Peters of Worcester who renamed her *Duchess of Worcester* and used her to travel and power his set of galloping horses. The rest of her working days were spent with this firm until she was scrapped in 1933. Regretfully no details are known about this delightful photograph but it could easily have been taken when Peters acquired the engine in 1908. What would be more natural than to pose the family with their new engine? *(courtesy Leeds Traction Engine Club)*

16. Two interesting photographs taken at Barnstable Fair circa 1908 during the erection of Hancocks' bioscope show. Clearly shown above is Burrell 8nhp compound engine No 2355, *Lord Roberts*, Registration No XH 8048. Built in March 1901, she was sold to W.C. & S. Hancock of Bristol for travelling his two wagon front walk up bioscope show. By 1919 she was with Fred Gray of Hampstead who had previous associations with the Hancocks, he renamed her *Mons Star*. She had changed owners twice more in showland by September 1941 when she was with J. Bond of Hampstead and by May 1943 she had been sold to J. Biddall also of Hampstead. Finally in 1947 she went to J.W. Hardwick & Sons of West Ewell for scrap. *(courtesy Rev. F. Larkworthy)*

18. Another view of Hancock's Bioscope Show, this time built up and open for business at Redruth Fair circa 1918. Burrell No 2355 *Lord Roberts* is shown at work towards the end or her service with Hancocks
*(courtesy D.E. Hattam)*

19. A collapsed bridge in Norfolk, circa 1908: the engine is Burrell 8nhp compound No 2456, *Edward VII*, Registration No. CM 3586, which was built in February 1902 for James Crighton of Norwich, who was her only owner in showland.
*(courtesy B.D. Stoyel)*

20. Burrell 7nhp single crank compound engine No 2462, *Alexandra*, Registration No WR 6979, built in March 1902 and sold to Charles Thurston & Sons of Norwich to travel their first bioscope show. This was known as Thurstons Royal Show, on one side of the show front was an 89 key Gavioli organ whilst on the other side was No 2462 partly hidden behind elaborate carvings and sign boards.
In this picture, taken at Oxford's St Giles Fair in 1906 the name plate *Alexandra* is quite clear, but the records state she was named *Queen Alexandria*.
*(courtesy the Hon W.H. McAlpine)*

21. Burrell 8nhp compound engine No 2497, *Excelsior*, built in August 1902 and sold to William Taylor of Ilkeston, Derby and shown here powering his Royal Bioscope Show at Oxford St Giles Fair in 1905. In April 1907 he bought another Burrell No 2898 and named it *Victoria* after one of his daughters. He then changed the name of *Excelsior* to that of his other daughter, *Alexandra*, presumably to keep the peace in the family. Sometime later he sold the engine to A. Twigdon of Leicester, her last owner. *(courtesy Hon W.A. McAlpine)*

22. Photographed circa 1913 at a fair in the London area is Burrell 8nhp compound engine No 2562, *Empress of the South*, Registration No HR 3896. She was built in April 1903 as a road locomotive and sold to J. Land & Sons of Portland, Dorset. Early in 1912 she went to R. Townsend & Sons of Weymouth, who converted her into a showmans engine and travelled her for four years before selling her to William Nichols of Forest Gate, Essex in 1916. In 1939 he sold her out of showland. *(author's collection)*

23. Shown here at Hampton Court Fair on the 3rd August 1935 is Burrell 8nhp compound engine No 2576, Registration No KE 5879. Built in June 1903 as a road locomotive she was sold to Herbert E. Blunt of Sholden, Kent and was later bought by A. Traylen of Earlsfield, London who converted her into a showmans engine and named her *Marina*. In June 1950 she was sold to J.W. Hardwick & Sons, of West Ewell, Surrey.
*(courtesy B.D. Stoyel)*

24. Burrell 8nhp compound engine No 2631 *King Edward VII*, Registration No SG 2500, built in November 1903 and sold to H.J. Scard (Wadbrook & Scard) for travelling their Bioscope Show. At the beginning of the Great War in 1914 she was sold to Henry O'Brien of Gateshead who gave her the name *Willing*. Later she went to R. Wylie & Sons of Haddington, East Lothian, who were to be her last owners. Photographed at Newcastle Town Moor Fair on 20th June 1936, her sideboards read 'Carnival Productions'.
*(courtesy B.D. Stoyel)*

25. Another view of Burrell No 2631, *King Edward VII*, here standing out of use at Chertsey on Saturday 13th May 1939.
*(courtesy A.C. Durrant)*

26. Burrell 7nhp compound engine No 2633 *Sir Redvers*, Registration No BP 6791, was built in November 1903 as a road locomotive and sold to Henry Arden of Heavitree, Devon. During the next twenty years she had three more owners until 1923 when she was sold to A. Traylen of Earlsfield, London who converted her into a showmans engine and renamed her *Shamrock*. By 1943 she was standing derelict at their Feltham yard. Shown here at Hampton Court Fair on 3rd August 1935. *(courtesy B.D. Stoyel)*

27. Photographed at Congleton Fair on 11th May 1906 is Burrell 8nhp compound No 2650, *Independent*, Registration No DH 2479. She was built in February 1904 and sold to Relph & Pedley of Northampton who renamed her *Premier*. In 1907 she was sold to Pat Collins of Bloxwich where she again underwent a name change now being called *Mark Twain*. In 1924 she was disposed of to John McGurk, a showman of Tramore in County Waterford. *(courtesy B.D. Stoyel)*

(courtesy the Hon W.H. McAlpine)

28 & 29. Two photographs of Burrell 8nhp compound engine No 2651■, *Challenger*, Registration No AY 9840, taken in 1907 en-route from Lincoln to Boston. She was built in March 1904 and sold to Mrs Annie Holland of Mile End, London. This engine was bought to replace the portable engine that powered her bioscope show that was known as Hollands Palace of Light. Later she was acquired by Charles Holland of Swadlincote, Derby and travelled his three abreast set of galloping horses roundabout until just before the First War. He sold her out of showland in the fifties.
*(Above)* washing the loads and *(below)* with four loads and a water dandy. The living van at the rear of the column is horse drawn.

(courtesy S.White)

30. Burrell 8nhp compound engine No 2668■ *Britannia*, Registration No NM 257, at Cambridge Mid Summer Fair on 2nd June 1935. She was built in June 1904 for William Thurston of Cambridge who later in the year formed a partnership with his brother Henry Thurston Junior to acquire a bioscope show. On one side of the entrance to the show was an 89 key Gavioli organ whilst on the right hand side stood *Britannia*. In 1911 the partnership was dissolved and William acquired both the engine and the show. By 1914 bioscope shows were finished and soon after the start of the war Thurston's Palace of Light, as the show was known, ceased to exist and by early 1920 *Britannia* was travelling a steam driven gondola ride acquired from Pat Collins and managed by Stanley Thurston, William's eldest son. In 1932 Stanley took over the engine and travelled their first set of dodgems which he had bought from Patsy North of Hunstanton. In 1946 she was sold to H. Gumble of Willingale, Essex and not long after that she went to her last showland owner Albert Manning of Cottered, Herts. *(courtesy R.G. Pratt)*

31. Another view of Burrell No 2668■, *Britannia*, this time driving Thurston's dodgems at Stotfold Feast in Bedfordshire on 16th July 1937. *(courtesy R. Smith)*

32. In a spot of trouble at Shrewsbury, circa 1927, is Burrell 8nhp compound engine No 2713, *His Majesty*. Built in December 1904, she was sold to Pat Collins of Bloxwich and travelled a bioscope show for many years until the show was taken off the road at the beginning of the First War. She then travelled the former Wall & Hammersley show which was converted to a menagerie and travelled under Mrs Flora Collins management up to the time of her death in 1933, when she was then laid aside. *(courtesy B.D. Stoyel)*

33. Posing for the official photograph in the works yard is Burrell 8nhp compound engine No 2721, *Edinburgh Castle*, Registration No SG 3125. Built in February 1905, she was sold to John Evans of Edinburgh for whom she worked for thirty-five years before going for scrap in 1940. *(author's collection)*

34. Photographed at Bridgewater Fair in September 1909 is Burrell 8nhp compound No 2723, *Doris,* Registration No HT 3468. Built in March 1905 she was sold to James Leo of Preston, Lancashire for travelling his bioscope show and then passed into the ownership of Marshall Hill of Bedminster, Bristol and renamed *Bristol City.* After spending a number of years driving their motor switchback she travelled various other of the organization's rides before being sold to Edward Harris, also of Bristol. She was finally scrapped in 1950.

*(courtesy R. Smith)*

35. A later photograph of Burrell No 2723, unfortunately the date is not known, but it is interesting to see the changes that have taken place. Gone is the Dickenson dynamo which has been replaced with a Mather & Platt of more conventional shape. The front valance has also been removed enabling the registration plate to be displayed on the front cross member, and solid rubber tyres have been fitted to the wheels. The Burrell plate advising that the engine is mounted on springs has gone and a manhole has been made in the end of the belly tank to facilitate cleaning the interior.

*(courtesy B.D. Stoyel)*

36. Driving Anderton & Rowlands show at a West Country fair is Burrell 8nhp compound engine No 2744, *Jumbo.* Registration No. HO 5577 and built in November 1905 as a road locomotive, she was sold to Henry Arden of Heavitree, Devon. In March 1907 she was sold to Anderton & Rowland of Bristol who converted her into a showmans engine and renamed her *Unique, and later Admiral Beresford.* In June 1916 she was sold out of showland to F. Houghton Ltd. of Durley, Hampshire. In 1931 she came back into showland on being sold to the firm of Wall Bros of Petersfield who renamed her *Admiral.* Eventually she was sold to Mrs Blackmore of Bideford, Devon who travelled her until she went for scrapping.

*(courtesy B.D. Stoyel)*

37. Providing power for Charles Thurston's Electric Cinema Palace is Burrell 8nhp compound engine No 2780■, *King Edward VII*, Registration No CL 4300. Built in November 1905 she was sold to Charles Thurston & Sons of Norwich, who travelled her with his Great Bioscope show. In 1943 she was sold to her last showland owner, Charles Presland of Tilbury, Essex. Note the very tall extension chimney designed to take the smoke well away from the front of the show in this photograph taken at Woodbridge in 1915. *(author's collection)*

38. Driver Pye Bridgeman and his mate Joe Walsh pose for this photograph of Burrell No 2780■, *King Edward VII*, whilst the engine provides the power for Thurston's Peacock Scenic Ride.

*(courtesy Leeds Traction Engine Club)*

39. Official works photograph, taken in Burrell's yard in 1906, of Burrell 4nhp compound tractor No 2802, Registration No AH 06. She was built in March of that year and sold to Lord John Sanger of Horley, Surrey for travelling his menagerie show in company with Burrell tractor 2803. In 1922 both engines were sold to Charles Hart of Barking, Essex but shortly after they were split up when sold to different owners and passed out of showland. These tractors were both fitted with direct steering by rack and pinion although they were both subsequently modified to normal chain steering. *(author's collection)*

40. In 1926 Charles Hart of Barking sold Burrell No 2802 to boilermakers J. Hickey & Son of Cheshunt, Hertfordshire. She is seen here standing derelict in Hickey's yard on 16th May 1936. *(courtesy B.D. Stoyel)*

41. Burrell compound engine No 2803, Registration No AH 07, was built in March 1906 and sold to Lord George Sanger of Horley, Sussex for travelling their menagerie in company with Burrell 2802. She performed this task until the early 1920s when she was sold to Charles Hart of Barking and after two further changes of ownership she finished her working life with Hayes (Pembroke Dock) Ltd. *(author's collection)*

42. Burrell 8nhp compound engine No 2804■, *The White Rose of York*, Registration No BR 1497, was built in April 1906 and sold to Alfred Payne of York for travelling with his four abreast set of gallopers. By 1923 she had been sold with the gallopers to Anderton & Rowland of Bristol and she drove from Sunderland to Babbacombe in Devon with the galloper loads. In 1929 she passed into the ownership of Charles Heal & Sons of Glastonbury, Somerset and in 1941 was sold to R. Wilson of Sunbury, Middlesex. Her next owner was Pat Collins of Bloxwich, Staffordshire who renamed her *The Griffin*, her registration number was changed to DH 2542 and she took the makers plate from No 2788. *(author's collection)*

43. Burrell 6nhp compound engine No 2806, *Sir Colin Campbell*, Registration No DB 2689, was built in May 1906 and sold to James Hibbert of Stockport for travelling his set of galloping horses. She spent all of her working life with this concern and became a familiar sight on fairgrounds in the North Western area. Her proud owner poses on the left of this photograph, taken in St Johns Road, Buxton, whilst travelling the galloper loads from Buxton Fair on 17th July 1906. *(courtesy B.D. Stoyel)*

44. Burrell 5 ton compound traction No 2876■, Registration No P 3093 (later changed to PB 9615), is shown at Basingstoke Carnival in 1937. She was built in January 1907 and sold to Thomas Mitchell & Sons of Guildford. Later in her life she was sold to amusement caterers Wall Bros of Petersfield who named her *Little Samson* although they did not fit a full length canopy or dynamo bracket. She was later sold out of showland. *(courtesy R. Smith)*

45. Photographed in 1938 arriving at Taunton with the dodgem loads is Burrell 8nhp compound engine No 2877■, *His Majesty*, Registration No HT 3163. Built in February 1907 she was sold to Charles Heal & Sons of Glastonbury, Somerset with whom she spent all of her working life. She was sold out of showland in 1959.

*(courtesy R. Smith)*

46. Burrell 8nhp compound engine No 2979, *Reliance*, Registration No BN 4883 was built in March 1908 and sold to Relph & Pedley of Northampton to travel and light their new set of flying pigs roundabout. On the 8th June, after only three months of work, she was involved in a bad accident at Thirlmere in which her driver was killed and she finished upside down in the lake. She was returned to Charles Burrell Ltd and rebuilt as No 3038, emerging in November 1909 whereupon whe was sold to Reuben Holdsworth of Keighley, who renamed her *The Prince* and continued using her with the flying pigs in association with Relph & Pedley. In 1936 whe was sold to Pat Collins of Bloxwich where she worked until being scrapped. Photographed at Nelson Fair on 20th July 1932 whilst in the ownership of Reuben Holdsworth.

*(courtesy R.D. Stoyel)*

47. A rather sad photograph of Burrell 8nhp compound engine No 2983, *Lady Violet*, lying derelict at New Brighton, Cheshire in 1930. Built in March 1908 and sold to James Crighton of Norwich for travelling his palace of light bioscope show, she spent the whole of her working life with this firm and was named after his daughter.

*(courtesy B.D. Stoyel)*

48. Taking water on the Great North Road, Tyneside, circa 1913, is Burrell compound tractor No 3028, Registration No AH 022. Built in September 1908, she was sold to Tory Bros of Sturminster Marshall in Dorset. Later she was bought by James Graven & Sons of Ely who converted her into a showmans tractor and sold her, in 1911, to Bostock & Wombwell who named her *Pedro*. *(courtesy the Hon W.H. McAlpine)*

49. Burrell 6nhp compound engine No 3063, *Houpla*, Registration No CR 5893 was built in February 1909 and sold to J.H. Herbert of Southampton with whom she travelled for the next twenty four years. The canopy side boards read 'J H Herbert Caterer of Amusements'. *(author's collection)*

50. A period photograph of Charles Thurston's 'Royal Bioscope Show' taken at Braintree Fair in 1910. Just visible on the right of the entrance and providing power for the show, is Burrell 8nhp compound engine No 3075■, *Alexandra*, Registration No CL 4301. She was built in February 1909 for Charles Thurston & Sons of Norwich for travelling with this enlarged show, taking the place of Burrell No 2462, also named *Alexandra*. Later she was acquired by John Thurston of Colchester, Essex who in 1949 sold her out of showland.

*(author's collection)*

51. Burrell No 3075■ *Alexandra*, photographed on this occasion providing power for the scenic motors switchback in company with Burrell No 3200■, *Unity* at Cambridge Fair in June 1910. *(courtesy B.D. Stoyel)*

52. John Green poses in front of Burrell 7nhp compound engine No 3089, at Kings Lynn Mart, circa 1930, at a time when the engine was travelling the caterpillar ride. Named *His Majesty* with Registration No CK 3347 she was built in March 1909 and sold to George Green of Glasgow. She was later acquired by his son John for travelling his Golden Dragon scenic ride in company with Burrell 3444, *His Lordship.* *(author's collection)*

53. Burrell 7nhp compound engine No 3159■, *Gladiator*, Registration No CO 3823 was built in December 1909 and exhibited that month at the Smithfield Show. When the show closed she was delivered to Anderton & Rowland at Pontypridd rather than their Bristol base. She was originally acquired to power their bioscope show but spent most of her life working the Venetian gondolas. In December 1930 she was sold to T. Whitelegg & Sons of Plymouth to haul and power their American radio cars ride. From 1941 to 1946 she was laid up at Launceston and then towed to their yard at Fisher Square, Exeter where she stood idle until she steamed away into preservation on 10th July 1953. *(courtesy D.E. Hattam)*

54. Burrell 7nhp compound engine No 3178, *Dreadnought*, Registration No AD 9000 was built in March 1910 and sold to Hill Bros of Bedminster, Bristol to travel their scenic ride. By September 1914 she had been sold to F.W. Nash of Cheltenham and Gloucester, a relative of Marshall Hill. In 1934 she was sold to T. Whitelegg & Sons of Plymouth who travelled her with their waltzer ride. She went to her last owner in 1945, J. Russett & Sons of Southampton used her for one season before she was scrapped. This is the official works photograph taken in March 1910 prior to delivery, by road judging by the coal stacked on the tender. *(author's collection)*

55. This photograph of Burrell No 3178, *Dreadnought* was taken in September 1936, twenty-six years later, at Bridgewater Fair whilst in the ownership of T. Whitelegg & Sons. By now the feed water pipe from the pump to the boiler clack valve above the belly tank had been removed and solid rubber tyres fitted, the original number plate had also been removed. *(courtesy J. Hobbs)*

56. Burrell 8nhp compound engine No 3200■, *Unity*, Registration No CL 4296 was built in May 1910 and sold to Charles Thurston & Sons of Norwich. She was bought to replace the Savage portable light engine which powered their scenic motors switchback ride and in her later years she travelled a set of dodgem cars. *Unity* is shown at Cambridge Fair on 6th June 1915, her driver H. Youll is in the centre. *(courtesy R. Smith)*

57. Burrell 6nhp compound No 3211, *The Dreadnought*, Registration No HO 5942 was built in January 1911 as a road locomotive and sold to W. Seward & Son of Petersfield, Hants. She passed through three changes of ownership before being bought by Wall Bros in 1935 who converted her to a showmans engine. She was finally sold for scrap to J.W. Hardwick & Sons of Ewell, Surrey in 1948. *(courtesy J. Hobbs)*

58. This pig roast at Stratford on Avon's Mop Fair in 1912 was held in Rother Street. In the background is Burrell 8nhp compound engine No 3272, *King George V*, Registration No AB 8921. She was built in March 1911 and sold to 'Uncle Tom' Clarke, as he was affectionately known, of Kidderminster, Worcestershire for driving his scenic motor car switchback along with sister engine No 3273. *(courtesy R. Smith)*

59. Fresh from the paint shop at Burrell's works is *Queen Mary*, Burrell 8nhp compound engine No 3273, Registration No AB 8922, the sister engine to *King George V* in the photograph above. Built in the same month and sold to the same owner she travelled the same switchback for many years. 'Uncle Tom' Clarke died in the mid-twenties and his widow continued to travel the ride until her death in 1931. The following year both engines and ride went to Uncle Tom's stepson Frank Wilson who travelled both engines until they were scrapped in 1949. *(author's collection)*

60. Where this unfortunate mishap occurred to Burrell No 3273, *Queen Mary*, is not known but it appears that she was unable to hold her loads whilst descending a hill. *(courtesy R.G. Pratt)*

61. Burrell No 3273, *Queen Mary*, preparing to leave Quinton Club, Gloucestershire with Frank Wilson's Noahs Ark loads. *(author's collection)*

62. Burrell 8nhp compound engine No 3277, *King George V*, Registration No HT 2915, was built in March 1911 and sold to Hill Bros of Bedminster, Bristol. She was to replace Burrell No 3178, *Dreadnought*, in travelling the second Savage-built scenic railway which they had taken delivery of in October of the previous year. The engine on the right of this photograph, which was taken at Chard on 14th January 1942, is 8nhp compound engine No 3866, *Vanguard*, Registration No HT 2914, which was built in December 1920 and sold to Marshall Hill of Bedminster.

(courtesy J. Hobbs)

63. This photograph of Burrell No 3277, *King George V*, shows the engine arriving at Taunton Fair in August 1937.

(courtesy J. Hobbs)

64. Shown against the background of Portobello Fair, Glasgow in 1934 is Burrell 6nhp compound engine *Baby the Flower of Belshill*, Registration No G 8413, built in February 1911 and sold to H. Harris of Glasgow. In 1921 she was bought by Dan Taylor, also of Glasgow, who shortened the name to *Baby* and travelled her with his swings and other rides until she was finally laid aside and scrapped in 1943. *(courtesy R. Smith)*

65. Pausing for the photographer at Olney, Bucks sometime between 1911 and 1920 is Burrell 8nhp compound engine No 3285 ■. Named *King George V* and given Registration No AH 5304 she was built in March 1911 and sold to Abbott & Barker of Norwich who travelled her with their motor car switchback for a number of years. In 1930 she was sold to Swales Forrest of Dartford, Kent for use with their gondola switchback. She was eventually sold out of showland in 1957. *(author's collection)*

66. Burrell 6nhp compound engine No 3343■, Registration No DX 3441, was built in October 1911 as a road locomotive and sold to Ralph Dumma & Son of Jedburgh, Roxburgh. She is shown at Ipswich Fair, held on the London Road Ground, on 8th July 1934 whilst in the ownership of Bert Stocks. He acquired the engine in June 1922 and converted her into a showmans engine and renamed her *Princess Mary*. She spent the rest of her showland days with this firm before being sold. (courtesy R.G. Pratt)

67. Burrell 8nhp compound engine No 3352, Registration No DH 2480, was built in December 1911 as a road locomotive and sold to Moore Bros of Bilborough, Notts. In May 1914 she was sold to Pat Collins of Bloxwich who converted her into a showmans engine and named her *Faugh a Bella*, she spent the rest of her working life with this firm. This photograph was taken at Birmingham Fair on 26th May 1934. (courtesy R. Smith)

68. Burrell 5 ton compound tractor No 3369, Registration No AH 090, built in May 1912 and sold to George Carter & Sons of West Southborne, Hants. Early in 1920 she was bought by Symond & Cook of Gloucester who converted her into a showmans tractor and named her *Gloucester City.* Later she carried a name plate *Little Edward* after their son who is seen sitting on the dynamo bracket in this picture taken at a Gloucestershire fair in the late twenties. In 1938 she was sold to Mrs R. Russell of Shepton Mallett, Somerset. Although the exact date is not recorded she was eventually scrapped by F. Moss of Leigh on Mendip.

*(author's collection)*

69. Burrell 8nhp compound engine No 3371, *Lord George,* Registration No BK 4931, built in March 1912 she was sold to Jacob Studt (senior) of Maesteg and later acquired by John J. Studt of Portsmouth. In 1927 she was purchased by J. Rose of Chertsey, Surrey who travelled her with his motor car scenic ride. In 1934 she went to Henry P. Studt of Reading and in 1942 to Sammy Mayne of Earlsfield. He replaced her front solid tyres with new wheels fitted with pneumatic tyres. In 1944 she was sold into Ireland. Photographed in Walpole Park, Ealing on Friday, 30th July 1939.

*(courtesy A.C. Durrant)*

70. Providing the power for Corrigans Speedway at Halifax Fair on 5th July 1940 is Burrell 8nhp compound engine No 3404, *Kathleen Mary,* Registration No WR 6763. Built in August 1912 and sold to Aspland & Howden of Boston, Lincs for travelling their bioscope show. When this show was finally taken off the road she was used with their gondolas. By July 1929 she was in the ownership of Edwin Corrigan of Dewsbury who used her to power his speedway ride. By 1942 she had once more changed owners and travelled for a short time with W. Noble of Consett, Co. Durham before being scrapped in the ownership of W. Bell of Stanley, Co Durham.

*(author's collection)*

71. Burrell 8nhp compound engine No 3441■, *William IV*, Registration No J 6858 was built in February 1913 and sold to William Murphy of Newcastle upon Tyne for travelling his scenic motor cars where she was later accompanied by Burrell 3610 *William V*. She was eventually acquired by Thomas Murphy. On the original photograph the words 'Scenic Railway' can just be discerned on the canopy sides. *(author's collection)*

72. In 1934 Burrell 3441, *William IV*, was sold to Captain Rowlands to travel his set of dodgems and she was driven down by road to join her new owner at Plymouth. By the time this photograph was taken at Babbacombe Regatta Fair on 13th August 1939 the turret crane had been removed. *(courtesy B.D. Stoyel)*

73. Celebrating the Coronation at Bolton Wakes Fair on 27th June 1953 is Burrell 8nhp compound engine No 3444■, *His Lordship*, Registration No CK 3403. Built in March 1913 and sold to Green Bros of Preston she spent most of her time in their ownership powering the golden dragons scenic ride. In December 1935 she was sold to Silcock Bros of Warrington who eventually sold her out of showland in 1949. *(B.D. Stoyel)*

74. Burrell 8nhp compound engine No 3450, *Always Ready*, Registration No J 6986, was built in March 1913 and sold to John Murphy of Newcastle upon Tyne. She was used to travel his scenic motor ride in company with Fowler 14515 *The Lion* before she was sold to A. Newsome of Darlington in 1925. Last licensed in 1948 she was finally cut up for scrap at Middleton St George, near Darlington, in 1950. *(author's collection)*

75. 'On His Majesty's Service', Burrell 6nhp compound engine No 3471 ■, *The Rover*, Registration No J 3471, at Barnards Castle after being requisitioned for war work in 1914. Built in May the previous year she was sold to A. Newsome of Darlington for travelling his 'Electrical Jumping Horses' roundabout. *(author's collection)*

76. Burrell 5nhp compound engine NO 3480, *Helpmate*, Registration No DW 2286. Built in June 1913 as a road locomotive and sold to Robert Wynn & Sons Ltd of Newport, Monmouthshire. By 1923 she had been bought by Alfred Smith of Bristol who converted her into a showmans engine and renamed her *Lord Lascelles*. She spent the rest of her working life with Reuben Gilham of Freemantle, Southampton, who acquired her in 1932 and this photograph was taken during that period at Southampton Fair on 22nd July 1942. *(author's collection)*

77. The scene at Newcastle upon Tyne during the Bostock & Wombwell auction on Monday 30th November 1931, when that firm's transport was sold. On the right standing above the crowd is Burrell 5nhp compound engine No 3509■, *Rajah*, Registration No G6469. She was built in November 1913 and sold to Bostock & Wombwell of Glasgow and named after one of the lions in their show. The tractor in the centre is Robey No 37660, *Bengal*, which travelled with the partnership during their last season in 1931. The Robey was sold for £25 and *Rajah* and *Nero* made £380. *(courtesy the Hon W.H. McAlpine)*

78. Burrell 7nhp compound engine No 3526■, *Lightning II*, Registration No AO 6302 was built in November 1913 and sold to Emerson & Hazard of Whitehaven, Cumberland for travelling their set of galloping horse roundabouts. She would appear to be still doing that work when this photograph was taken at Settle Fair on 20th June 1936. *(courtesy B.D. Stoyel)*

79. Burrell 6nhp compound engine No 3542, Registration No NO 472 was built in January 1914 as a road locomotive and sold to James Penna & Sons of Ruthern Bridge, Cornwall. In April 1916 she was sold to John Sadd & Sons Ltd, of Maldon, Essex. In 1929 she was bought by Charles Openshaw Ltd of Reading, converted into a showmans engine and sold the same year to James Matthews of West Worthing, Sussex who named her *Sunny South.* In 1947 she was sold to J. Beach of Sunbury, Middlesex with whom she worked for the next ten years before being scrapped. Photographed at Ealing Common Fair on the 31st May 1947.
*(courtesy A.C. Durrant)*

80. Photographed at Newcastle Town Moor Fair in 1925 is Burrell 8nhp compound engine No 3547, *Amy,* Registration No FT 643. Built in February 1914 and sold to Walter Murphy of North Shields for travelling his steam scenic motors ride. After the First World War she travelled their newly acquired Orton & Spooner electric scenic motor car ride. Several years later when the motor cars had lost their novelty value they were replaced with dolphins.
*(author's collection)*

81. Another view of Burrell 3547 which was renamed *Tucker* in September 1935. Before being acquired by John Powell of Jarrow, Co Durham. The location is again Newcastle Town Moor Fair this time in June 1938.
*(author's collection)*

82. Burrell 5nhp compound engine No 3555■, *The Busy Bee*, Registration No AO 6262 was built in March 1914 and sold to Taylor Bros of Workington, Cumberland who travelled her with their 'Unique Alpine Motors' ride. She was sold out of showland in 1940. This unfortunate accident occurred when the roof of an old cellar collapsed under the weight of the engine at Waverton, near Wigton, Cumberland, in 1929. *(author's collection)*

83. The accident to Burrell 3555■, *Busy Bee*, shown from the opposite (flywheel) side. *(author's collection)*

84. An impressive line up of engines all belonging to A. Danter & Sons of Newport, Monmouthshire, at Newport Fair on 10th August 1933. Left to right: Burrell 8nhp compound No 3590, *Kitchener*, Registration No NC 2250, built in June 1914; Burrell 8nhp compound No 3561, *Fearnought*, Registration No DW 2200, built in April 1914; Fowler R2 compound No 11824, *Invincible*, Registration No DW 2201, built in March 1910; and Burrell 8nhp compound No 3936, *Renown*, Registration No DW 2833, built in July 1922. *(courtesy R.G. Pratt)*

85. Burrell 8nhp compound engine No 3610■, *William V*, Registration No J 6857 was built in August 1914 and sold to William Murphy of Newcastle upon Tyne. For a number of years she travelled his scenic motors ride in company with Burrell 3441, *William IV*. In 1933 she was sold to Samuel N. Ingram of Hyde, Cheshire and later to Bason & Sons of Longsight, Manchester. She was acquired by James Botton & Sons of Green Street, Green, Kent and in 1955 she was sold out of showland. This photograph was taken at Queens Road Fair, Manchester on Saturday 5th August 1940. *(courtesy A.C. Durrant)*

86. Norbiton Fair on 22nd June 1935 is the setting for this photograph of Burrell 8nhp compound engine No 3659, *Lord Kitchener*, Registration No XC 9891. She was built in May 1915 and sold to O'Brien Bros of Winchester with whom she spent the whole of her working life. At first she travelled their gondola switchback ride but later was with their dragons scenic. In 1947 she was sold for scrap to J.W. Hardwick & Sons of West Ewell, Surrey. *(courtesy B.D. Stoyel)*

87. Burrell 6nhp compound engine No 3660, *Victor*, Registration No HT 3244 was built in May 1915 and sold to James Jennings of Upper Studley, Wilts. By 1919 she was in the ownership of David Harrison of Bristol, she was then acquired by F. Oldridge of Plymouth who sold her to J. Rowland & Sons of St Austell, Cornwall with whom she stayed until she was finally scrapped in 1950. With the drivers patiently waiting, she is seen taking water from a roadside ditch on 26th June 1936. *(courtesy R.G. Pratt)*

88. Burrell 5nhp compound showmans engine No 3669■, *Nero*, Registration No G 6470 was built in July 1915 and sold to E.H. Bostock of Glasgow. Like her sister, Burrell 3509■ *Rajah*, she too was named after one of the lions in the show. The last performance of their menagerie took place on 1st December 1931 at the Old Cattle Market, Newcastle upon Tyne, the day after their transport had been sold at auction. Note the rubber cross strakes which are easily seen in this undated photograph. *(author's collection)*

89. Burrell 5 ton tractor No 3693, Registration No AH 0267 was built in October 1915 and sold to George Lewis of Reading and in 1921 she received the new Registration No KE 3162. Her fourth and last owner was amusement caterer William Penfold of Plumstead, Kent who named her *Topsy*, and it was during his ownership that this photograph was taken — probably in the early thirties. *(author's collection)*

90. Pulling the galloper loads at Nelson Park, Hounslow, Middlesex on 18th May 1948 is Burrell 8nhp compound engine No 3694, Registration No MD 6145. She was built in October 1915 as a road locomotive and sold to Bath & Portland Stone Firms Ltd. She was later sold in the early twenties to William Beach Senr of Oxbridge, Middlesex who converted her into a showmans engine and named her *Lord Fisher*. In December 1948 his daughter Sally Beach sold her to J. Hardwick & Sons of West Ewell in part exchange for Burrell No 2894, transferring the name to the new engine. *(courtesy A.C. Durrant)*

91. Burrell 8nhp compound engine No 3724, *Victory*, Registration No AY 9836 was built in July 1916 and sold to A. Holland of Swadlincote, Derbyshire. She spent all of her working life with this firm most of it driving their scenic motor car ride. Note how the extension chimney is permanently hinged to the top of the chimney in this photograph taken at Northampton Fair in the early twenties. *(courtesy the Hon W.H. McAlpine)*

92. Photographed at Hounslow Fair on the 10th April 1939 is Burrell 5 ton compound tractor No 3786■ *Tiger*, Registration No PB 9610. Built in June 1918 and sold to E. Longhurst & Sons Ltd of Epsom, Surrey. In 1935 she was bought by J. Beach of Sunbury on Thames who converted her to a showmans tractor and renamed her *Sunny South*. Eventually owned by James Beach Junior of Hounslow, Middlesex who sold her to Joseph Mayne of Hounslow in May 1942.
*(courtesy A.C. Durrant)*

93. Burrell 8nhp compound engine No 3787, *Thomas William*, Registration No WR 6641, built in July 1918 for Goldthorpe Marshall Jnr of Bradford and named after Willie Marshall. She travelled their scenic motor cars, this ride was originally a steam driven gondola switchback but was modernised in 1906 by replacing the gondolas with the motor cars which were a novelty at that time. In 1930 she was sold to Walter Shaw & Sons of Sheffield for driving their electric speedway. They renamed her *Her Majesty*. In 1939 she was bought by Wall Bros of Petersfield, Hants being sold to J.W. Hardwick & Sons, the well known scrap dealers of West Ewell, Surrey in 1952. This photograph was taken at Millfield Cattle Market, Skipton on the 23rd May 1934.
*(courtesy B.D. Stoyel)*

94. Burrell No 3787 generating for Shaw's electric speedway at Hellifield Fair circa 1936.
*(courtesy B.D. Stoyel)*

95. A not uncommon sight after the Second World War when showmen were abandoning steam in favour of petrol and diesel war surplus vehicles, this photograph was taken in the yard of E.R. Goodey at Twyford, Berkshire on 26th December 1951. Right: Burrell 7nhp compound engine No 3804■, *Independence*, Registration No BP 5539, was built in May 1919 as a road locomotive and sold to W.T. Heaver of Ratham, West Sussex. In 1923 she was sold to Woolls Bros of Erith, Kent and converted into a showmans engine.

Centre: Burrell 8nhp compound engine No 2701■, *Black Prince*, Registration No PE 8171 was built in September 1904 and sold to J.W. Harris & Son of Hook Norton, Oxon.

Left: Foster 8nhp compound engine No 3663, *King of Showman*, Registration No EM 1176, was built in February 1909 and sold to Pat Collins of Bloxwich, Staffs. *(courtesy J.H. Meredith)*

96. Burrell 7nhp compound engine No 3810, Registration No CL 4317 was built in June 1919 and sold to Edward J. Edwards of Norwich. In 1923 she was sold to John Evans of Brecon who converted her into a showmans engine and in 1925 sold her to Hill Bros of Bedminster, Bristol, who named her *Hercules* and used her to travel their scenic ride. Marshall Hill died in February 1927 and his widow travelled the engine for that season, selling her with the cakewalk in 1928 to G. Rogers & Sons of Chipping Sodbury, Gloucestershire who renamed her *Princess Elizabeth*. In July 1942 she was sold to Dan Baker of Hook, in Hampshire who worked her with his three abreast gallopers. In 1944 she was acquired by Charles Evan & Sons of Newport, Monmouthshire who sold her, in 1945, to Conrad Studt, her last owner, before being scrapped. Photographed at Frome, Somerset in 1926. *(author's collection)*

97. Burrell 7nhp compound engine No 3817, *Perseverance*, Registration No MA 5741, at Colne Fair on 31st October 1933. Built in September 1919 and sold to Arthur Bates of Rhode Heath, Cheshire, with whom she spent the whole of her working life.
*(courtesy B.D. Stoyel)*

98. A gear side view of Burrell No 3817, *Perseverance*, taken at Colne Fair on 12th October 1939. She was scrapped in 1943.
*(courtesy B.D. Stoyel)*

99. Burrell 8nhp compound engine No 3827■, *Victory*, Registration No CL 4299, built in May 1920 and sold to Charles Thurston & Sons of Norwich for travelling his golden dragons scenic ride in company with Burrell 2780■ *King Edward VII*.
*(courtesy B.D. Stoyel)*

100. Burrell 8nhp compound engine No 3833■, *Queen Mary*, Registration No BR 1498 was built in May 1920 and sold to F. Payne of York for use with his four abreast set of gallopers. Around 1922 this ride, the 89 key Marenghi organ, Burrell 2804, *The White Rose of York* and *Queen Mary* were acquired by Anderton & Rowland of Bristol and they were driven down from Sunderland to join their new owners at Babbacombe Fair. In 1943 *Queen Mary* was sold to Mrs Symonds of Gloucester who was to be her last owner in showland. Photographed driving Anderton & Rowlands dodgems at Exeter Easter Fair in 1940. *(courtesy J. Hobbs)*

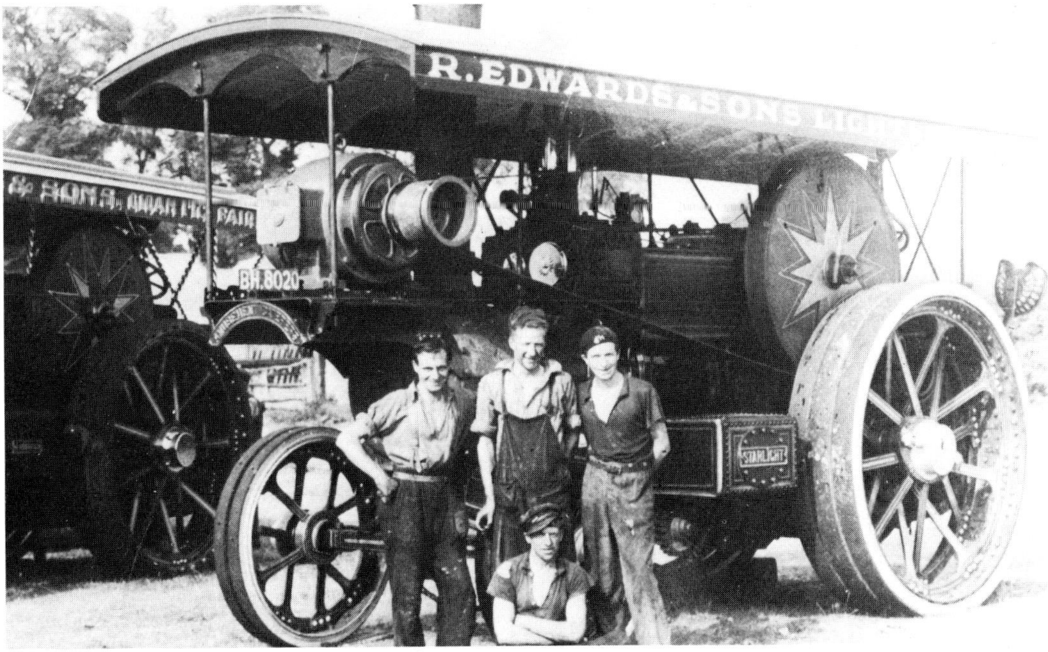


101. Burrell 6nhp compound engine No 3836■, *Starlight*, Registration No BH 8020 was built in July 1920 as a road locomotive and sold to Noah Judd of Chackmore in Buckinghamshire. She had two more owners before she was bought in 1934 by R. Edwards & Sons of Swindon who converted her into a showmans engine. She was sold out of showland after the Second World War. Precisely when this happy group of drivers posed for the camera is not known but it was immediately after the signwriters had finished the side boards as their chalk lines are still clearly visible on the original photograph. *(courtesy Leeds Traction Engine Club)*

102. Burrell 6nhp compound engine No 3847■, Registration No FB 3035 was built in July 1920 as a road locomotive and sold to Richard Drake & Co of Sutton, Ely. In 1921 she was sold to Mrs Hannah Parkin of Norwich who converted her into a showmans engine and named her *Princess Marina*, she also received a new Registration No CL 4483. Photographed on the road at Capel St Mary, near Ispwich in the mid-nineteen thirties, by this time she was in the ownership of John Barker & Sons of Ipswich who acquired the engine in 1933 and who was her last showland owner. *(courtesy R.G. Pratt)*

103. Left: Burrell 8nhp compound engine No 3865■, *No 1*, Registration No DH 2507, built in November 1920, and right: McLaren 10nhp compound engine No 1623■, *Goliath*, Registration No DH 2482, built in June 1918; photographed side by side at Aston Fair, Birmingham on 7th June 1952 when both engines belonged to Pat Collins of Bloxwich. *(courtesy J.H. Meredith)*

104. Burrell 8nhp compound engine No 3865■, *No 1,* Registration No DH 2507 was built in November 1920 and sold to Pat Collins of Bloxwich for travelling his Scenic Whales ride. Eventually she was employed on their Noah's Ark ride for several years and had the distinction of being the last scenic type Burrell remaining at work. Photographed at Leicester Fair held on Lee Street car park in October 1950. *(courtesy E. Woods)*

105. The gearside detail of Burrell No 3865■, *No 1,* is well illustrated in this photograph taken at the Birmingham Onion Fair on 29th September 1951 — note the onions hanging from the canopy sides in honour of the event. *(courtesy B.D. Stoyel)*

106. Another early fifties view of Burrell No 3865■, *No 1*, this time at Chester Fair in May 1951.

*(courtesy F.Jones)*

107. Burrell 8nhp compound engine No 3866, *Vanguard*, Registration No HT 2914 was built in December 1920 and sold to Marshall Hill of Bedminster, Bristol for travelling his motor scenic ride. She was later acquired by Bernard Hill of Leominster who scrapped her in 1952. *Vanguard* is seen at Frome Fair circa 1929.

*(courtesy J. Hobbs)*

108. Burrell 7nhp compound engine No 3871■, *Teresa*, Registration No DX 3076 was built in March 1921 and sold to Sidney Stocks of Ipswich who later changed her name to *Pioneer*. In 1930 she was sold to Hardiman & Strong of Bristol, who traded as John Cole, and they renamed her *Western Pioneer*. She is seen here powering their American Speedway ride at Mangotsfield Fair on Whit Monday 1940. By November 1955 she had been sold out of showland. *(courtesy R. Smith)*

109. Burrell 5 ton wagon No 3883, *Electra*, Registration No CL 4325 was built in February 1921 and sold to C. Summers of Norwich. By March 1925 she had been purchased by Smith Bros of Plymouth and was last used by them in 1934. *Electra* stood in their yard at Plymouth until destroyed in an air raid during the last war. She is seen here driving fully loaded down Gunnislake Hill, Cornwall in August 1932. *(author's collection)*

110. Standing derelict in the yard of J.W. Hardwick & Sons of West Ewell in February 1952 is Burrell 8nhp compound engine No 3886■, *Lord Lascelles.* Built in April 1921 she was given the Registration No XF 8162 and sold to Harry Gray of Battersea, London where she spent all of her working life until sold to Hardwicks in 1951.
*(courtesy F. Jones)*

111. Burrell 8nhp compound engine No 3888■, *General Gough,* Registration No NO 2379 was built in June 1921 and sold to Swales Bolesworth of Dagenham, Essex for travelling his gondola switchback ride. She is pictured standing disused whilst in the ownership of D.W. White of Margaretting in Essex, who acquired the engine in later years. Barely discernible on her, by now, well weathered side boards is the legend 'British Amusements on Tour'.
*(courtesy F. Jones)*

112. A line up of drivers and fairground workers pose for the camera in front of Burrell No 3896■, *Earl Beatty* at Exeter Easter Fair held by the canal at Avon Banks in 1934.
Right to left: Bert Saunders, Ernie Austin, Arthur Davies, Garry Whitehead, Carlo Mireno, George Brewer and Jack Maggs. *Earl Beatty's* driver, Dick Davies, is concealed in the shadows of the cab. (courtesy R. Davies)

113. Burrell 8nhp compound engine No 3896■, *Earl Beatty*, Registration No CO 4052 was built in May 1921 and sold to Anderton & Rowland of Bristol. She left the works at Thetford and steamed to Orton & Spooner's works at Burton-on-Trent to await completion of Anderton & Rowland's new super scenic railway. She departed from Burton-on-Trent on 23rd July hauling several loads and arrived at Newton Abbot in Devon on the 27th, the remaining loads following later by train. She spent the whole of her working life with Anderton & Rowland being sold out of showland in 1948. This photograph shows her driving the jungle speedway at Vivary Park, Taunton in May 1937. (courtesy J. Hobbs)

114. Burrell 5nhp compound engine No 3898, *Little Joe*, Registration No OH 5647 was built in March 1921 and sold to Joseph Fletcher of Birmingham. She travelled with him until 1935 when she was acquired by Christopher Thompson of Keighley. In 1937 she went to John Barker & Sons of Ipswich and then to James Severs of Eastington Loftus before passing to her last owner Sam Crow & Sons of Middlesbrough who scrapped her in 1948. She is seen here at Bradford Fair held in Harold Park in June 1934. *(author's collection)*

115. Burrell No 3898, *Little Joe* hauling her loads at Capel St Mary, Ipswich on 1st September 1939 when in the ownership of John Barker & Sons of Ispwich. *(courtesy R.G. Pratt)*

116. Burrell 8nhp compound engine No 3912■, *Dragon*, Registration No CO 4485 was built in November 1921 and sold to Anderton & Rowland of Bristol, and is seen in this photograph driving their diving dolphins ride at Paignton Regatta Fair in 1934. When new this engine was reputed to be the first to be fitted with pressed on rubber tyres to her rear wheels. In the winter of 1932/33 she went into Fowlers Works for an overhaul and when she emerged she was minus the pulley and top fittings of the turret crane. In 1943 she was bought by Sam Smart & Sons of Bedminster, Bristol. Five years later she was sold out of showland.

*(courtesy R. Davies)*

117. Burrell No 3912■, *Dragon* powering Anderton & Rowland's Brooklyn Racers in September 1938 at Rockwell Green Fair — note the water dandy which is just visible on the left of the photograph.

*(courtesy B.D. Stoyel)*

118. Burrell No 3912■, *Dragon* forms the background to this interesting photograph which was taken on Burrell's stand at the Royal Show at Derby in 1921. The Royal Show is usually held in July and this would indicate that the engine was returned to the works as her date of building was November 1921. On the left of the group is Colonel Wilberforce Burrell, Mr Burge of A.G.E., Herbert J. Burrell and Sidney O. Harrison, Burrell's sales representative.

*(author's collection)*

119. This birds-eye view of a steam fair at work was taken at Penzance Corpus Christi Fair in 1925. There are ten engines visible in this line-up, all but one at work; clearly illustrated in the foreground are three of Anderton & Rowland's engines hard at work generating power for their respective rides. Nearest the camera is Burrell 8nhp compound No 3912■, *Dragon*, Registration No CO 4485, built in November 1921; and behind her is Burrell 8nhp compound No 3896■, *Earl Beatty*, Registration No CO 4052, built in May 1921.

*(courtesy D.H. Hattam)*

120. Burrell 7nhp compound engine No 3945, *Prince of Wales*, Registration No HT 6804 was built in November 1922 and sold to Harry Coneley of Bristol. After eleven years with this firm she was sold to William H. Marshall & Sons of Bradford who travelled her until 1946. She is photographed at Gisburn, Lancashire on 24th August 1935 while in Marshall's ownership. *(courtesy B.D. Stoyel)*

121. Burrell 8nhp compound engine No 3949■, *Princess Mary*, Registration No NO 8287, was built in February 1923 and sold to William Nichol's of Forest Gate, Essex. Around 1944 she was sold to Charles Presland of Tilbury, Essex who travelled her until 1958 when they sold her out of showland. This is Burrell's official works photograph taken prior to the engine being sent to London for the 1922 Smithfield Show, almost two months before her official date of building. *(author's collection)*

122. Charles Presland poses alongside Burrell No 3949■, *Princess Mary* at Braintree Fair in 1949.

*(author's collection)*

123. This photograph of Burrell 8nhp compound engine No 3969, *Majestic*, Registration No WT 2255, was taken at Huddersfield on 12th April 1941. She was built in February 1924 and sold to Enoch Farrar of Doncaster for use with his Golden Dragon scenic ride. She was later acquired by John Farrar of Masborough in the West Riding of Yorkshire who travelled her until 1943.

*(courtesy B.D. Stoyel)*

124. Burrell 8nhp compound engine No 4021, *Lord Curzon*, Registration No CH 4829, was built in April 1925 and sold to Harry Hall of Derby. She was later sold to Pat Collins of Bloxwich where she finished her working life. The Burrell on the right of the picture is 8nhp compound No 3968, *The Whale*, Registration No CH 4111, which was built in February 1924 and also sold to Harry Hall, with whom she spent the whole of her working life. Both these engines travelled Hall's well known Scenic Whales switchback ride. Driver Tommy Glover stands by *Lord Curzon's* front wheel. *(author's collection)*

125. Burrell 8nhp compound engine No 4030■, *The Dolphin*, Registration No FA 2316, was built in September 1925 and sold to William S. Davies of Stoke-on-Trent who named her after the Diving Dolphins scenic ride that she travelled. In February 1937 she was sold to J. Shaw & Sons of Sheffield who renamed her *The Guvnor*. She is seen in Shaw's livery at Newcastle Town Moor Fair on 14th June 1937.

*(courtesy B.D. Stoyel)*

126. His Royal Highness the Duke of York posing alongside a Burrell showmans engine belonging to Barker & Thurston on the occasion of his visit to Savages works at Kings Lynn in January 1922. The gentleman on the left is the Mayor of Kings Lynn, in the centre is the late John Pilling, managing director of Savages Ltd.

*(author's collection)*

127. Photographed travelling to Banbury Fair in 1934 is Burrell 8nph compound engine No 4092, *Simplicity*, Registration No EU 4439. She was the last showmans engine to be built bearing the Burrell name, in October 1930. She was sold to Mrs Alf Deakin of Brynmawr in Brecon and was delivered to her at the Stratford Mop Fair that year. Most of her working life in showland was spent travelling a set of dodgem cars. In the late thirties she was sold out of showland.

*(courtesy R. Smith)*

128. Clayton & Shuttleworth 5 ton wagon No 47250, Registration No BD 5605 which was built in August 1915 and sold to the War Department. After the First World War she was purchased by W. Hailstone of Yarwell in Northamptonshire and by February 1921 she had been sold to amusement caterer Sam Smart & Sons of Bedminster, Bristol. Her last owner was also a showman, E. Andrews of Tunbridge Wells, Kent. She was last licenced in August 1929.
*(author's collection)*

129. A Clayton & Shuttleworth 4nhp tractor which was travelling with Baker & Gunnetts Wild Australia show. Unfortunately no details are available of the engine number, neither are Baker & Gunnett recorded as ever owning a Clayton tractor.
*(author's collection)*

130. Clayton & Shuttleworth 5 ton compound wagon No 48709, Registration No DO 1982, built in February 1921 and sold to Ben Howden & Sons of Boston, Holland where she spent the whole of her short working life, by 1928 she had been laid aside!
*(author's collection)*

131. Photographed at Hazel Grove, Cheshire in July 1910 is Foden 6nhp compound engine No 380, *Pride of Cheshire*, Registration No NE 113, built in August 1898 and sold to Edmund Holland of Macclesfield who used her for travelling his electric galloping horses roundabout.
*(courtesy B.D. Stoyel)*

132. Foden 8nhp compound engine No 528, Registration No BP 6589, she was built in July 1902 as a road locomotive and sold to the Royal Engineers at Chatham. By 1910 she had been bought by William Beach of Hammersmith who converted her into a showmans engine and named her *George V*. Shown here at Lindfield Fair, Sussex on 12 August 1939 in the ownership of Tom Smith & Sons of Shoreham by Sea, West Sussex to whom she was sold at the beginning of January 1921 and with whom she worked until being scrapped in July 1949.
*(author's collection)*

133. Foden 6nhp compound engine No 1540, *Prince Llewellyn*, Registration No FE 4204, built in March 1908 and sold to Mrs Simons & Son of Caernarvon and Leicester. She later travelled with Hill Bros of Bedminster, Bristol who sold her to Charles Ling of Doncaster who renamed her *Duchess of York*.
*(courtesy R. Smith)*

134 & 135. Official works photographs of Foden 6nhp compound engine No 380, *Pride of Cheshire,* Registration No NE 113, built in August 1898 and sold to Edmund Holland of Macclesfield for travelling and powering his electric galloping horses roundabout. She had seven more owners, the last being James Coppenhall of Sandbach in 1947, although it is not clear if he was a showman. The decorative finish of the engine is of interest, as is the uncovered spoked flywheel. Most moving parts were covered as much as possible to avoid frightening passing horses. *(author's collection)*

136. Photographed at Holbeck Fairground, Leeds on 10th December 1910 is Foden 8nhp compound engine No 2104■, *Prospector,* Registration No WR 6985. Built in March 1910 she was sold to Walter Shaw & Sons of Sheffield and used to travel their steam swing boats. In 1956 she was sold out of showland.
*(courtesy B.D. Stoyel)*

137. The same engine twenty-four years later; Foden No 2104■, *Prospector,* at Stump Cross, Brierfield on 29th May 1934. By this time she had been fitted with solid rubber bolt-on tyre sections.
*(courtesy B.D. Stoyel)*

138. Foden 4 ton wagon No 5606, *Lord Kitchener,* Registration No M 7523, built in September 1915 and sold to Ernest Ashcroft, amusement caterer, of Barnsley whose son eventually joined him in the business. She spent the whole of her working life with this firm and was last licensed in October 1930 and by November 1933 she had been scrapped.
*(courtesy Leeds T.E. Club)*

139. Foden 4 ton wagon No 6512, Registration No M 8870, was built in November 1916 and sold to Henry Kingham & Sons Ltd of Watford, Herts. After the war she was sold to amusement caterers Woolls Bros, of Erith, Kent who named her *Prince*. In 1926 she was acquired by F. Smith & Sons of Belvedere, Kent. Her last owner was Job Martin, also from Belvedere, who bought her in 1937. *(author's collection)*

140. Foden 3 ton wagon No 9324, Registration No MA 1633 was built in October 1919 and sold to the Manchester Brewery Co Ltd. She is seen here at Bolton Fair in the early thirties whilst in the ownership of Richard Cubbins, a showman of Rochdale and her fourth owner, he bought the wagon in 1930 and travelled her with his swings and side stalls. The date of scrapping is not recorded. *(courtesy B.D. Stoyel)*

141. Foden 6 ton wagon No 11400, *Worcester City*, Registration No FK 2063 was built in April 1924 as a showmans wagon for A.J. Peters of Worcester. He was her only owner during her working life and she was finally cut up by Flower & Sons of Stratford upon Avon. *(author's collection)*

142. Foden tractor No 12372, Registration No NR 9505, was built in September 1926 and sold to Coltman Bros of Claybrooke Magna in Leicestershire. In 1929 she was sold to Simons & Greatorex of Nantwich, Cheshire who converted her into a showmans tractor and named her *Little John*. This official works photograph taken at Sandbach in 1929 after the conversion shows the engine and galloper loads. *(author's collection)*

143. Foden 6 ton wagon No 13028, Registration No BA 7403 was built in July 1928 and sold to the Bleachers Association Ltd, Salford, Lancashire. Eventually she was bought by M.A. Collins of Hightown, Manchester and named *Michael*. She was to spend the rest of her working life with this firm. In this photograph taken at Burnley Fair on 9th July 1935 only the extension chimney on the top of the cab distinguishes her as a showmans vehicle, the dynamo being mounted on the body and the driving belts passing through holes cut in the back of the cab. *(courtesy B.D. Stoyel)*

144. The end of the road for Foster 8nhp compound engine No 3053, *Victor*, Registration No VK 6782, which was built in July 1905 and sold to Robert Carr of Sunderland. In March 1908 she was sold to Alf Twigdon of Dunton Bassett, Leicestershire and then had a succession of owners: John Monk of Leicester in 1914, J.J. Wallis of Bootle, J.E. Whiting of Sheffield, S. Hadfield of Southport and finally, in 1931, T. Miller & Son of Willington Quay, Northumberland. In July 1935 she ran out of control at Denton Bank, Newcastle, turned over and had to be cut up on the spot. *(author's collection)*

145. Foster 8nhp compound engine No 3642■, *Maude*, Registration No PB 9832, was built in August 1908 as a road locomotive and sold to C.F. Rowley of Burslem, Staffordshire. On 21st June 1911 she was sold to J. Rose of Chertsey, Surrey who converted her into a showmans engine. Since this photograph was taken, on 24th March 1951, she has been restored and is still travelled by the firm. *(courtesy J.H. Meredith)*

146. Foster 8nhp compound engine No 3663, *King of Showmen*, Registration No EM 1176 was built in February 1909 and sold to Pat Collins of Bloxwich, Staffordshire. In August 1913 she was sold to W.T. Smith of Leicester and Lincoln and renamed *Sanspareil*. In December 1916 she came south in the ownership of W. Gritt of Romsey, Hants who re-named her *Nellie* before selling her to G.H. Wallis of Bootle, Lancashire. She came south again in 1929 when bought by Woolls Bros of Erith, Kent who were destined to be her last owners. She is seen in Woolls' ownership at Catford Hospital Fair on 24th July 1933. *(courtesy R. Smith)*

147. Foster 8nhp compound engine No 12463, *Roosevelt*, Registration No NC 2251, photographed at Harpurley Fair, Manchester on 15th April 1941. She was built in December 1909 and sold to John Collings & Son of Liverpool for travelling and lighting his four abreast galloping horse roundabout. She was eventually acquired by J.P. Collins of Manchester with whom she worked out the rest of her days.
*(author's collection)*

148. With the galloper loads, circa 1920, is Foster 8nhp compound engine No 12499, *Marvel*, Registration No HO 5702. Built in May 1910 she was first sold to William Irvin of Fleet, Hampshire and was later acquired by George J. Irvin of Earlsfield, Surrey with whom she spent the rest of her working life. Prior to purchasing this engine William Irvin had suffered some lean times but by 1910 he was in a position to trade in his old Burrell No 1655, *William the Conqueror*, as part payment on the new engine so named because he said 'It is a marvel that we have been able to buy such a fine engine after so many ups and downs'.
*(author's collection)*

149. Foster 8nhp compound engine No 12504, *King George V*, Registration No XC 9908, built in May 1910 and sold to H.J. Wallis of Seaforth, Lancs. By the early twenties she had been sold to Dagnalls Ltd of Willesden Green, Middlesex and by 1921 was travelling with Woolls Bros of Erith, Kent when she was photographed with canopy sideboards reading 'Amusements of Distinction'.
*(courtesy B.D. Stoyel)*

150. Foster 8nhp compound engine No 12517, *Queen Mary*, Registration No CB 3063, built in June 1910 and sold to E. Davies of Stoke on Trent. By June 1917 she was travelling in the ownership of Jack Whyatt of Stoke on Trent who sold her to E Collins of Castleford in 1943. Her last owner was R. Wilson of Sunbury, Middlesex who scrapped her in 1951. Photographed at Nelson Fair in Lancashire on the 17th July 1934. One canopy side board reads J. Whyatt whilst the other side spells his christian name out in full.
*(courtesy D.B. Stoyel)*

151. Another view of Foster No 12517, *Queen Mary*, in typical surroundings at the back of the fair details of this photograph are not recorded.
*(courtesy S. White)*

152. Foster No 12517, *Queen Mary*, using her feast crane to help build up the ride at Blackburn Fair on the 14th April 1938.
*(courtesy B.D. Stoyel)*

153. Foster 6nhp compound engine No 12538, *Surprise*, Registration No NK 1745 was built in October 1910 and sold to Mrs Catherine Bird of Watford, Herts. In 1929 she was sold to John Flanagan of Watford and stayed with him for the rest of her working life. For a short time during the First War she was used by the army for transporting horse fodder and was engaged in this task at Haddenham railway sidings near Oxford when this photograph was taken in 1915. *(courtesy the Hon W.H. McAlpine)*

154. Driver Reg Brice poses in front of Foster No 12538, *Surprise*, at Boreham Wood Fair in 1930. *(courtesy Leeds Traction Engine Club)*

155. Foster 6nhp compound engine No 12590 was built in May 1911 and sold to Sarah & Johnson & Son of Sheffield. By 1931 she was travelling with Horne Bros of Dundee and had been given the name *Flower of the Spen*. Eventually she was sold to S. Littler of Bryn, Wigan but it is not known how many years he travelled her. This sad photograph was taken at Littler's yard at Bryn Hall, Wigan in 1953. *(courtesy B.D. Stoyel)*

156. Foster 65bhp compound engine No 12620, *Prince Bert*, Registration No AY 9838 was built in July 1911 and sold to Holland Bros of London and Swadlincote, Derbyshire. She was named after Bert Holland, one of six sons. Seen here at Swadlincote Fair in 1912, the engine spent her entire working life with this concern. *(Leeds Traction Engine Club)*

157. Photographed at Nelson Fair, Lancashire on 17th July 1940, is Foster 65bhp compound engine No 12940, *Lancashire Lad*, Registration No TB 3704. She was built in August 1912 and sold to William Mitchell & Sons of Heywood, Lancashire with whom she spent the whole of her working life. *(courtesy B.D. Stoyel)*

158. Foster 6nhp compound engine No 13010, *Cock o' the North*, Registration No SG 3363, seen at Penrith Whitsun Fair on 16th May 1932. Built in May 1913 and sold to W. Bastable & Sons of Stirling, she had two more changes of ownership during her working life, being with W. & C. Forrest of Bonnybridge, Stirling and William Ingram of Kirkcaldy although the dates are not recorded. *(author's collection)*

159. Foster 8nhp compound engine No 13020, *Prince of Wales*, Registration No BU 1907, was built in March 1913 and sold to Hughes & Martin of Liverpool. By 1925 she had been acquired by William Martin, also of Liverpool. In 1937 she was sold to Toogood & Jobson and changed owners again in 1953 when she was sold to William Murphy of Jarrow. She is seen here at Queen's Road Fair, Harpurley, Easter 1929.

*(Leeds Traction Engine Club)*

160. Foster 5nhp compound engine No 13034, *Little Gem*, Registration No J 6785, was built in February 1914 and sold to Luke Rich of Newcastle on Tyne. She travelled his rides and side stalls until September 1939 when she was sold to A. & G. Taylor of Asselby, in the East Riding of Yorkshire. By 1943 she was travelling with Edwin Knowles of Braithwell and in 1947 was sold to her last owner Joshua Rodgers of South Crosland in the West Riding. This photograph was taken at Newcastle Town Moor Fair circa 1931.

*(author's collection)*

161. Another early thirties scene at Newcastle's Town Moor Fair shows Foster 65bhp compound engine No 13052, *No 1*, Registration No DH 2477. Built in April 1913 she was sold to Pat Collins of Bloxwich who used her to travel his No 1 scenic railway. In 1935 the engine was renamed *Clara* after his second wife whom he married that year, and who had managed the scenic railway from its beginning. The engine was acquired by John Collins Senior of Liverpool and then by M.A. Collins of Manchester with whom she spent the rest of her working life. Just visible on the canopy sides in the original photograph are the words 'It's Collins's Fair'. *(author's collection)*

162. Foster Wellington compound tractor No 14202, Registration No HO 5723, was built in November 1915 and sold to Gardam & Co of Staines, Middlesex. By February 1921 she had been sold to showman Henry Chipperfield of Fleet, Hampshire who named her *Helpmate*. She was only used for haulage and was never fitted with a dynamo and by 1930 had been sold to Ernest Barry of Dublin, who was to be her last owner. *(author's collection)*

163. Shown here providing power for the swirl at the Cambridge Mid-Summer Fair on 26th June 1935 is Foster 7nhp compound engine No 14153■, Registration No BE 7221. Built as a road locomotive in July 1916 and sold to the Farmers Co of Brigg, Lindsey she passed into the ownership of William Thurston & Son of Cambridge in February 1922 who converted her into a showmans engine and named her *Admiral Beatty* and used her with the Golden Gondola switchback. During the twenties she was taken over by Stanley Thurston and in 1932 she travelled their gallopers until these were sold a year later. The firm then took delivery of a new Lakin-built swirl which was driven by *Admiral Beatty*. In 1948 she was sold out of showland.
*(courtesy R.G. Pratt)*

164. Foster 7nhp compound No 14388, Registration No AH 6265 was built as a road locomotive in September 1920 and sold to Aldritch Bros Ltd of Diss, Norfolk. In 1927 she was bought by J. Harris & Sons of Biggleswade, Beds and converted into a showmans engine and given the name *Surprise* and for a number of years she travelled their set of galloping horses. She remained with this firm until scrapped circa 1952.
*(courtesy B.D. Stoyel)*

165. Photographed in 1941 at Uxbridge Fair is Foster 7nhp compound engine No 14381, Registration No CL 4318, built in October 1919 as a road locomotive and sold to Edward J. Edwards of Norwich. By 1927 she was owned by engine dealer George Taylor of Redbourn, Herts who converted her into a showmans engine and sold her to A. Traylen of Earlsfield, London who named her, *Queen Mary*. She was later sold to J.W. Hardwick & Sons of West Ewell who scrapped her.
*(author's collection)*

166. Dewsbury Fair on 21st July 1941 is the setting for this photograph of Foster 65bhp compound engine No 14402, *Admiral Beatty*, Registration No NL 1431. Built in December 1919 she was first sold to Dagnells Ltd of Willesden Green, London and then followed a succession of well known showland owners, J. Hoadley of Newcastle upon Tyne, by 1922 William Whyatt of Jarrow, then John Proctor of Belper, by 1931 J. Pullen of Sheffield and by 1933 E. Farrar of Doncaster who renamed her *Junior*.
*(author's collection)*

167. Shown here in the early twenties in Victoria Railway Station, Sheffield, is Foster 65bhp compound engine No 14403, *James Walter*, Registration No WA 3807. She was built in December 1919 and sold to J.E. Whiting of Owlerton, Sheffield for driving their scenic railway. Over the next five years she changed owners no fewer than three times before being bought by W.H. Marshall & Sons of Bradford in early 1934. She travelled their waltzer ride for the next few years being sold in August 1944 to J.C. Gray of Gateshead. By 1945 she was with F. Manders of Gateshead who were to be her last owners before she was cut up.
*(author's collection)*

168. Foster 65bhp compound engine No 14404, *Gladys*, Registration No WR 6758 at Lincoln Fair in April 1933. She was built in March 1920 and sold to E.C. Farrar Junior of Doncaster for use with his scenic golden dragons which made their first appearance at the Mid-Lent Fairs in the Eastern Counties in 1920. She was accompanied by Fowler 14845, *Reliance*, and spent the whole of her working life with this concern.
*(author's collection)*

169. Foster 65bhp compound engine No 14405■, *William Henry,* Registration No WA 5519 was built in October 1921 and sold to J.E. Whiting of Owlerton, Sheffield. In March 1927 she was sold to Pat Collins of Bloxwich, Staffordshire and her works number was altered to 14446, She was renamed *The Leader* and given a new Registration No DH 4593. In 1963 they sold her out of showland. This photograph was taken at Worcester Fair in 1932.

*(courtesy R. Smith)*

170. Foster 65bhp compound engine No 14446■ *The Leader* at Dudley Fair on 3rd June 1951.

*(courtesy J.H. Meredith)*

171. Foster 65bhp compound engine No 14446■*The Leader* in the works yard of Hunt Bros (Oldbury)Ltd on 17th July 1952, after receiving running repairs. *(courtesy W.T. Hunt)*

172. Foster 7nhp compound engine No 14500, *Cheshire Lad*, Registration No MA 5270 (later changed to MA 5271), was built in July 1920 and sold to J.H. Royle of Gorton, Manchester. In 1928 she was bought by H. Clayton of Liverpool and by the beginning of 1935 she was travelling in the ownership of W. Bastable of Stirling, who had renamed her *Ben Hur*. In 1945 she was bought by W.C. Forrest of Bonnybridge, Stirling and a year later she was sold to Andrew Wilson of Auchterdarran, Fife, her last owner. Seen here at Town Moor Fair, Newcastle on 14th June 1937. *(courtesy B.D. Stoyel)*

173. Foster 7nhp compound engine No 14502■, *Victory*, Registration No XC 9862. Built in August 1920 and sold to Mrs Catherine Bird of Watford. By the time this photograph was taken at Witney Feast in 1934 she was in the ownership of John Flanagan, also of Watford, who had had her canopy sides lettered 'Flanagan's Great Fun Carnival'. She spent the rest of her showland days with this concern. *(courtesy R. Smith)*

174. Foster 7nhp compound engine No 14503, *Prince of Wales*, Registration No KB 8952 was built in July 1921 and sold to John Ryan of Moreton, Cheshire. In 1935 she was sold to T. Drakeley of Stechford, Birmingham with whom whe spent the rest of her working life. Seen here at Moreton Fair on 7th July 1930. *(courtesy B.D. Stoyel)*

175. The two photographs on this page are of the same two engines taken from different sides. Above Foster 65bhp compound engine No 14404, *Gladys*, Registration No WR 6758, was built in March 1920 and sold to E.C. Farrar Junior of Doncaster with whom she spent the whole of her working life. *(courtesy B.D. Stoyel)*

176. Foster 65bhp compound engine No 14607, *Dornorber*, Registration No ET 6184, was built in April 1930 and sold to Enoch Farrar Junior of Doncaster with whom she spent her whole working life. She was named after Enoch and Gladys' three daughters, Dorothy, Norah and Bertha. When delivered she was painted black and lined out in red to match Enoch's Riley car. These photographs were taken at Kings Lynn Fair on 9th February 1932 whilst travelling their owner's Krazy Kars ride. *(courtesy B.D. Stoyel)*

177. Fowler 8nhp single cylinder traction engine No 3012, *Lightning I,* was built in July 1877 and sold to George Robson of Newtown, St Boswells, Roxburgh. By August 1906 she was in the ownership of Emerson & Hazard of Whitehaven. She is seen travelling their leaping horses roundabout at Windermere in 1906.

*(author's collection)*

178. Fowler A4/R compound engine No 7323, *Enterprise,* was built in October 1895 and sold to Tom Norman of Dallington, East Sussex. She travelled for only a short time with this firm and apparently did not come up to their expectations as she was returned to the works and is then recorded as being sold to an owner in Kildrum. Co Kerry.

*(author's collection)*

179. Fowler B4 compound engine No 7411, *Belle of the Road*, was built in August 1895 and sold to John Murphy & Walter Hoadley of Newcastle upon Tyne for travelling their four abreast set of galloping horses roundabout. She later became the sole property of Walter Hoadley with whom she spent the rest of her working life. This accident happened on 21st July 1920 after she had run back down a hill at Eltherley, Bishop Auckland, County Durham and turned over. *(author's collection)*

180. The official works photograph of Fowler B4 compound engine No 7412, *Powerful*, which was built in August 1895 and sold to William Murphy and J. & F. Newsomo of Newcastle upon Tyne. For just under two years she travelled their gondala switchback before being sold out of showland in November 1897. *(author's collection)*

181. Fowler A4R compound engine No 8331, *Electric*, Registration No KE 3307, was built in April 1899 and sold to Arthur Martin of South Wigston, Leicestershire who used her to travel his gondola switchback. By July 1918 she had been sold to Dagnalls Ltd of Cricklewood. Early in 1921 she was sold to J. Williams & Son of Cudham, Kent, before going to her last owner Fred Ward of Eltham, Kent. For a number of years she travelled and provided power for his set of dodgem cars. *(courtesy B.D. Stoyel)*

182. Fowler No 8331 *Electric* arriving in Hucknall High Street with Arthur Martins gondola switchback loads, seven wagons and a living van, circa 1899. *(courtesy the Hon W.H. McAlpine)*

183. Fowler B5 compound engine No 8725, *General Buller*, Registration No WR 6787, was built in April 1901 and sold to Jacob Studt of Maesteg and Newport, and used to travel his big show. In December 1904 she was sold to William H. Marshall of Bradford with whom she worked out her life until she was scrapped in 1947. She is seen at Fowler's works having undergone her steam trial and the fitting of her canopy, she is about to be pushed into the paintshop for completion. *(author's collection)*

184. Fowler No 8725 *General Buller* over thirty years later working at a village fair in the mid-thirties. *(author's collection)*

185. Fowler B5 compound engine No 8924, *Coronation*, Registration No HT 3078, was built in May 1902 and sold to John Studt of Cardiff. She arrived by train from Leeds and was steamed up and driven to Sophia Gardens accompanied by great crowds of people to take part in the fair that was open for King Edward's Coronation festivities. In January 1909 she was sold to George Pruett of Bristol and renamed *King of the Road*. In due course she was sold to Mrs Green of Congresbury, Somerset in whose yard she is seen. She is thought to have been standing for some time as she was reported there as early as September 1935.
*(courtesy B.D. Stoyel)*

186. Fowler A4/R compound engine No 9273, *Pride of Yapton*, Registration No HO 5753, was built in February 1902 as a road locomotive and sold to John Sparks of Yapton, West Sussex. By 1921 she had been acquired by Wall Bros of Petersfield, Hants. renamed *Independence* and converted to a showmans engine. In 1933 she went north to Scotland after being bought by Horne Bros of Dundee, Angus and she is pictured, on 10th July of the following year, at Kirkcaldy Fair which was held at The Basin. *(courtesy R. Smith)*

187. Fowler B6 compound engine No 9295, Registration No CU 913 was built in December 1902 and sold to the War Department, one of ten built with long fireboxes fitted for oil burning. By 1929 she had been sold to Maurice Stokes of Basingstoke, Hants who fitted a dynamo on a conventional bracket mounted on the smokebox but retained the original short canopy. He named her *Old Bill* and she is seen here at Hurstbourne Prior Fair on the 31st July 1935.
*(courtesy B.D. Stoyel)*

188. Fowler B5 compound engine No 9381■, Registration No WR 6770, built in April 1902 as a road locomotive and sold to John Smith's Tadcaster Brewery Co Ltd. By 1921 she had been sold to Christopher Johnson & Sons of Featherstone who converted her into a showmans engine and named her *Lady Betty*. Photographed at Heath Common Fair, Wakefield on 12th April 1934.
*(author's collection)*

189. Official works photograph taken prior to delivery of Fowler A5 compound engine No 9386■, *The Maid of Cefn Yddfa*, Registration No AX 2857. She was built in March 1904 and sold to William Haggar & Sons of Aberdare, Glamorgan for travelling their bioscope show which was the first of the Welsh shows to have a big organ front, a Marenghi — which used to play such tunes as 'The Village Blacksmith', 'I've got a Bungalow' and 'We Parted on the Shore' — complete with a troupe of dancing girls. The name of the engine was derived from a famous Welsh story a film of which was made by the Haggar family and often shown in both the Haggar shows. In 1914 whe was sold to her last showland owner J.M. Butlin who was in association with Sydney Squires and they renamed her *Lord Nelson*.
*(author's collection)*

190. Fowler A4/R compound engine No 9456■, Registration No KK 3634 was built in August 1902 as a road locomotive and sold to Robert Cann of Wareham, Dorset. She had at least four more owners before being sold to Tom Smith of Shoreham by Sea, Sussex, who converted her into a showmans engine and named her *Marina*. Photographed at Haywards Heath Fair on 27th July 1947, she was later sold out of showland.

*(courtesy B.D. Stoyel)*

191. Fowler B5 compound engine No 9487, Registration No HR 4186, was built as a road locomotive and sold to the War Department in January 1904. The engine had a varied career: she was acquired by R. Edwards of Swindon who converted her into a showmans engine; by 1931 she was in the ownership of Miles Jervis of Chasetown, Staffordshire who named her *Duke of York;* and by 1934 she was travelling with John Ryan of Moreton, Cheshire. Within a few years she had been bought by James P. Collins of Hightown, Cheshire and drove various rides belonging to this firm until she was scrapped in 1945. Shown in Collins' ownership at Sefton Park Fair, Liverpool on 12th May 1937 in company with Foster 12463 *Roosevelt*. *(courtesy B.D. Stoyel)*

192. Shown at Gloucester Fair in 1919 is Fowler B6 compound engine No 9502, *Emperor*, Registration No L 9317. She was built in March 1905 and sold to Jacob Studt of Maesteg who later renamed her *King Edward*. She was delivered to him at Reading and travelled his bioscope show for a number of years until this show ceased to travel when she was put to work with his galloping horse roundabout. In the early part of 1921 she was acquired by Henry P. Studt of Reading for whom she worked until scrapped in 1950.
*(courtesy R. Smith)*

193. Fowler B5 compound engine No 9817, *Progress*, Registration No KE 2309, built in September 1906 as a road locomotive and sold to Joseph Bassett of Trentham, Staffs. Later she was acquired by E. Hambrook & Sons of Swingfield, Kent, followed by Thomas Wood & Sons of Crockenhill, Kent. Her next owner was J.W. Waddington of Norwich who converted her into a showmans engine and worked her with his steam yachts in company with McLaren 829 also named *Progress*. Circa 1935 the engine was sold to Arthur Bird of Costessey Norfolk; she then passed into the ownership of J. Christian of Moordown, Hants with whom she spent the rest of her working life, and for many years she travelled his eight car dodgem track. Shown here at work with J.W. Waddington before he had fitted her canopy.
*(courtesy B.D. Stoyel)*

194. This photograph of Fowler No 9817, *Progress*, shows her with the canopy fitted at Wareham Fair on 14th September 1937.
*(courtesy B.D. Stoyel)*

195. Just visible behind the crowd and the display boards on the right of this pre-First World War picture is Fowler A4 compound engine No 9887, *The Mexican*, Registration No BL 8500. She was built in March 1904 and sold to Biddall Bros of Hounslow, Middlesex for travelling their Mexican Circus. By 1921 the engine had been sold out of showland. *(author's collection)*

196. Fowler D3 compound engine No 9961, Registration No XC 9890 was built in March 1904 as a road locomotive and sold to the War Department. After they had finished with her she was bought by J. Sanders of Lambeth who converted her into a showmans engine and named her *Surprise*. She spent the rest of her working life with this firm. Photographed on Hampstead Heath in the early 1920s. *(courtesy Leeds Traction Engine Club)*

197. Fowler B6 compound engine No 9974, *The Showman,* Registration No NL 5500, was built in December 1904 and sold to Randall M. Williams of West Hartlepool, the canopy side boards proclaiming 'Randall Williams Great Show No 1'. On April 9th 1909 she was sold to Samuel Stokes of Overton, Hants, as it was said she was too heavy for the Lancashire roads. She was noted at Basingstoke Fair in 1915, newly painted and having a scroll on the belly tank saying 'Who'd a thought it, horses bought it' which was appropriate as besides being a roundabout proprietor Sam Stokes was a horse dealer and a good judge of horses. By 1923 she had been sold to John Evans of Edinburgh, and then in early 1927 she was sold to her last owners T. Miller of Newcastle upon Tyne who worked her until after the Second World War. Seen here at the Crane Fair on 4th April 1938. *(courtesy B.D. Stoyell)*

198. A photograph believed to have been taken in Cardiff just after the First World War of Fowler B6 compound engine No 10323, *Admiral Togo,* Registration No HB 1638. She was built in July 1905 and sold to John Studt of Cardiff, who travelled her mainly with his four abreast gallopers. Mr Studt who was known as 'Big John' was an outstanding figure in Welsh showland. He possessed a very powerful voice and it was not unusual to see hundreds of people assembled in front of his switchback, from the steps of which he would sing the Welsh national anthem and other songs accompanied by the strains of the organ. In 1917 she was sold along with the four abreast to White Bros of Cardiff who renamed her *The Welsh Horse.* She spent the rest of her working life with this firm before being scrapped in 1938. *(author's collection)*

199. Fowler R1 compound engine NO 10693, *The Sun*, Registration No MA 5770, was built in June 1906 and sold to Captain Thomas Payne of Ashton-under-Lyne, Lancashire for travelling his Royal Bioscope show, which was fronted with a magnificent Limonaire organ. Early in 1925 she was sold to Timothy Ray of Mansfield Notts. Within a few years of the end of the last war she was sold to Samson Barker of Pentrich, Derby: she was scrapped in 1954. This official works photograph shows the unusual method of lowering the dynamo by means of pivoting the front awning supports. She was fitted with polished aluminum boiler and cylinder lagging. *(author's collection)*

200. Fowler R1 compound engine No 10696, *George the First*, Registration No HR 5373 was built in September 1906 and sold to Aspland & Howden of Boston, Lincolnshire for use with their bioscope show. At the end of the First World War she was sold to Dagnells Ltd of Cricklewood, London who resold her, early in 1920, to E. Hill & Sons of Swindon with whom she spent the rest of her working life. *George the First* is seen providing power for Aspland & Howden's bioscope show at Lincoln Fair circa 1912 as the proprietors and staff pose for the camera. *(courtesy the Hon W.H. McAlpine)*

201. Fowler A5 compound engine No 11107, *Monarch*, Registration No WR 6986, was built in April 1909 and sold to A. Wildman of Caernarvon. In 1913 she was sold to John Evans of Edinburgh. By January 1921 she was with Walter Shaw & Sons of Sheffield. In May 1939 she was sold to W.H. Church & Sons of Halifax who used her until she was scrapped in 1947. Seen here powering Shaw's new auto scooter cars at Newcastle Town Moor Fair circa 1937. *(author's collection)*

202. Photographed at Burnley Fair, Lancashire, circa 1930, is Fowler R2 compound engine No 11126, *Our Fancy*, Registration No NC 2509. Built in May 1908 she was sold to Farrar & Tyler of Coalville, Leicestershire for travelling their No 2 show. In 1912 she was sold to Joseph Ling of Doncaster and by 1919 was with Tippler White who renamed her *Felix*. In 1925 she was bought by John Collins Senior of Liverpool. She passed to J.P. Collins of Hightown, Manchester and then to Michael A. Collins before being sold for scrap in 1950. *(author's collection)*

203. Fowler A4/R compound engine No 11416, Registration No MA 5134 was built in March 1908 as a road locomotive and sold to John Wood of Crockenhill, Kent. She had two more owners before she was sold to H. Barnard of Rugby in 1924. He converted her into a showmans engine and named her *Esther*. After a few years she was sold out of showland and was scrapped in the late forties. The owner and his family pose in front of their engine at Alum Rock Fair, Birmingham in August 1925; second from the right is the driver, Bob Neal, who drove for Pat Collins for a number of years. *(courtesy R. Smith)*

204. Fowler R1 compound engine No 11815, *King George V*, Registration No CY 4307 was built in June 1910 and sold to Walter Haggar of Aberdare, Glamorgan and used by him to travel his picture palace. The canopy sides were lettered 'Haggar's Picture Palace'. During the First World War the show was laid up and they established a permanent cinema in Pembroke. Later she was sold to J.T. Danter of Pontypridd, Glamorgan, finally being scrapped in 1946. *(author's collection)*

205. Fowler B6 compound engine No 11846, *Lord James*, Registration No DH 2484, was built in 1909 and sold to J. & R.M. Williams of West Hartlepool, Co Durham and named after Randell Monte Williams' second son. All his sons had Lord as a christian name as the result of a suggestion made to him by Lord George Sanger. By 1919 she had been sold to Pat Collins of Bloxwich, Staffordshire and was used for driving their scenic whales switchback. Seen here at Burnley Fair on 11th July 1933. *(courtesy B.D. Stoyel)*

206. Fowler A7 compound engine No 12702, Registration No AD 9702 was built in May 1912 as a road locomotive and sold to Edmund Cullimore of Thornbury, Gloucestershire. In 1938 she was sold to Smart & Sons of Bedminster, Bristol for use with their four abreast set of galloping horses roundabout. She remained a plain road locomotive, never receiving showmans fittings, and the power was provided from a dynamo mounted on the back of a Crossley tractor which she drove with a belt off her flywheel. Photographed at Bridgewater Fair in 1944. *(courtesy J. Hobbs)*

207. Fowler B6 compound engine No 12910, Registration No WR 7149 at Town Moor Fair, Newcastle on 19th June 1937. She was built in April 1913 as a road locomotive and sold to G. George Gunter of Wetherby in the West Riding of Yorkshire. Early in 1930 she was sold to Charles Openshaw of Reading and converted into a showmans engine. The same year she was sold to J.W. Hoadley of Newcastle upon Tyne who named her *Jubilee*, and for a number of years she travelled his Skid Ride. *(courtesy B.D. Stoyel)*

208. Fowler B6 compound engine No 13047, Registration No DP 3863 was built in November 1914 as a road locomotive and sold to the War Department. After the war she was sold to Charles Openshaw of Reading who converted her into a showmans engine and sold her to J. Matthews of West Worthing who named her *Wanderer*. Circa 1948 she was sold to James D. Foster of Emsworth. She was cut up for scrap by Pollock Brown & Co Ltd of Northam, Southampton in the mid fifties. *(courtesy B.D. Stoyel)*

209. Fowler A7 compound engine No 13091, Registration No MA 5742, was built in October 1913 and sold to War Hoare of Bovey Tracey, Devon. She had three more owners before being sold to Morley Bros of Castleford in Yorkshire who converted her into a showmans engine and named her *Our Lizzie*. In 1947 she was sold to Steel & Son of Ardsley, near Barnsley who last used her in 1952. The cab and smokebox fitted when this photograph was taken at Holbeck Feast, Leeds, in 1934 came from Morley's old Foden engine.

*(courtesy B.D. Stoyel)*

210. The engine nearest the camera is Fowler A7 compound engine No 13149, Registration No KE 7014, which was built in July 1913 as a road locomotive and sold to W. Vousden & Son of Eastry, Kent. She had two more changes of ownership before she was sold in 1929 to J. Day, amusement caterer, of Edenbridge, Kent who converted her into a showmans engine. Later she was bought by Edward Pettigrove of East Peckham, Kent who was her last recorded owner. The middle engine of the trio is Fowler 14426 and the photograph was taken at Shoreham Fair in 1936 when both engines were in the ownership of J. Day.

*(courtesy R. Smith)*

215. Fowler B6 compound engine No 14110, Registration No AT 6356, was built in December 1914 as a road locomotive and sold to the War Department. By 1921 she had been sold to Albert Richards of Hull who converted into a showmans engine. Some years later she was sold to Goldthorpe Marshall Junior of Bradford who named her *Excelsior* and travelled her until the outbreak of the Second War. She is seen in Marshall's ownership at Burnley Fair held on the Cattle Market on 11th July 1931. *(courtesy B.D. Stoyel)*

216. Fowler 11410 *Excelsior*, seen in this photograph at Nelson Fair on the 19th July 1933.
*(courtesy B.D. Stoyel)*

217. Driver Bob Neal (right) poses at Dudley Fair on 28th May 1950 in front of Fowler B6 compound engine No 14424■, Registration No DH 2545 which was built in May 1916 as a road locomotive and sold to the War Department. After the war she was sold to Charles Openshaw of Reading who were well known for their showmans engine conversions. She was then sold as a showmans engine to Pat Collins of Bloxwich, who named her *Dreadnought*. In the mid-fifties she was sold out of showland. *(courtesy B.D. Stoyel)*

218. Fowler compound engine No 14425■, Registration No DP 4418 was built in May 1916 as a road locomotive and sold to the War Department. Circa 1924 she was sold to Charles Openshaw of Reading who converted her into a showmans engine and, in 1925, sold her to Frank Codona of Glasgow who named her *Carry On*. In 1944 she was sold to Kevin McGivern of Warrenpoint, Co Down. K.M. McDowell of Newtonards, Co Antrim bought her in 1954 and by 1957 had sold her to J. Ellis & Son of County Kildare who sold her out of showland in 1961. Seen here at work building up McGivern's three abreast galloping horses. *(author's collection)*

219. The Town Moor Fair, Newcastle in 1924 with the golden dragons switchback and Fowler B6 compound showmans engine No 14515, *The Lion*, Registration No SG 2498, built in May 1915 and sold to T.B. Murphy of South Shields and used with their steam driven motorcar roundabout for a number of years. In 1919 she was sold to John Evans of Edinburgh and put to work with his golden dragons switchback. In 1930 she went to William Murphy of Newcastle upon Tyne and was eventually acquired by John Powell of Jarrow, Co. Durham who was Mrs Murphy's nephew. For a number of years she travelled his waltzer and later his set of dodgem cars. She was retired from work in 1945.
*(courtesy B.D. Stoyel)*

220. Fowler No 14515, *The Lion*, poses with her driver.
*(author's collection)*

221. Photographed with the waltzer on 14th June 1937 at Newcastle Town Moor Fair is Fowler B6 compound 14515, *The Lion*.
*(courtesy B.D. Stoyel)*

222. Fowler B6 compound engine No 14845, *Reliance*, Registration No WR 6759, built in June 1920 and sold to Enoch Farrar of Doncaster. She travelled the firms scenic golden dragons in company with Foster 14404, *Gladys* and spent all her working life with this firm until sold for scrapping in 1945. Shown here at Newcastle Town Moor Fair on 13th June 1938.
*(courtesy D.B. Stoyel)*

223. Fowler R3 compound engine No 14863, Registration No FX 6689, built in March 1917 as a road locomotive and sold to the Ministry of Munitions. By 1921 she had been bought by F.J. Barnes of Portland Dorset and by 1933 had passed into the ownership of Jenning Bros of Devizes who converted her into a showmans engine and named her *Excelsior*. In 1938 she was sold to H. Symonds of London who renamed her *Shamrock* and she spent the rest of her working life with this firm. Shown here at a fair in the London area with H. Symonds' speedway cars.
*(courtesy B.D. Stoyel)*

224. Photographed at Taunton Fair in 1939 is Fowler R3 compound engine No 14879, Registration No MR 2241 which was built in April 1917 and sold to the Ministry of Munitions. After the war she was sold to Mark Hill & Sons of Warminster, Wilts who used her until 1936 when she was purchased by Hill Bros of Bedminster, Bristol and converted into a showmans engine with the name *King Edward VII*. She had two more owners before being exported in September 1951.
*(courtesy R. Smith)*

225. Fowler TE2 compound engine No 14916, Registration No NO 1246, built in September 1917 as a road locomotive and sold to the Ministry of Munitions. At the end of the war she went to G.E. Millbank of Hatfield Heath, Essex and in the early thirties she was sold to Fred Gray of the Vale of Health, Hampstead who converted her into a showmans engine and named her *Demolicious'* (although the date when she was so named is not clear it would be most probably circa 1941 when she was employed pulling down bombed buildings in London). She was later sold to J. Biddall of Hounslow, Middlesex before going to J.W. Hardwick & Sons of West Ewall for scrapping. *Demolocious* is seen here in Fred Gray's yard in Hampstead.
*(courtesy A.C. Durrant)*

226. Fowler R3 compound engine No 15116■, *Bertha*, Registration No CB 3062, built in April 1920 and sold to John Whyatt of Stoke on Trent. In 1943 she was purchased by Mrs J. Beach of Maidenhead, Berkshire and sold out of showland in 1956. Photographed at Burnley Fair, which was held on the Cattle Market, in July 1931.
*(courtesy B.D. Stoyel)*

227. Standing derelict at Maidenhead, circa 1954, is Fowler No 15116■, *Bertha*.
*(courtesy S. White)*

228. Fowler R3 compound engine No 15316, Registration No BL 813, at the fair in Liscard Park on 7th July 1936. Built in November 1918 as a road locomotive she was sold to the War Department and after the war she was eventually bought by Charles Openshaw and converted into a showmans engine. She was then sold to Frederick Wallis of Seaforth, Lancs who named her *Duke of York*.

*(courtesy B.D. Stoyel)*

229. Chard May Fair in 1938 is the setting for this view of Fowler R3 compound engine No 15319■, Registration No FX 7850. She was built in January 1919 and sold to the War Department but had several owners before being bought in the early thirties by R. Townsend & Sons who converted her into a showmans engine and named her *Queen Mary*. This firm was her only showland owner.

*(courtesy R. Smith)*

230. Providing power for Collins' swirl at Sefton Park, Liverpool is Fowler A9 compound No 15372, Registration No BD 5465. Built in 1919 as a road locomotive she was sold to W. Ivens & Sons of Welton, Northants and at some later date she was purchased by Northants County Council who sold her in 1935 to J. Corrigan & Sons of Dewsbury in the West Riding. She was sent into Fowlers works and converted into a showmans engine receiving the name *Jubilee*. The following year she was sold to E. Collins of Castleford and was to have two more owners before being scrapped in 1947. An old Tilling Stevens converted bus is parked alongside and the photograph was taken on 12th May 1937.

*(courtesy B.D. Stoyel)*

231. Fowler A9 compound engine No 15374, Registration No WR 8399 at Brierfield Fair on 21st May 1933. Built in October 1919 as a road locomotive she was sold to James Turner & Son Ltd of Kiveton Park, Sheffield. In 1929 she was sold to D. & N. Connelly of Blackpool who converted her into a showmans engine and named her *Peace*.
*(courtesy B.D. Stoyel)*

232. Fowler R3 compound engine No 15663, *The Mascot*, Registration No DP 3511, built in May 1930 and sold to Charles Openshaw of Reading. In February 1922 they sold her to H. Powell of Newcastle upon Tyne who travelled her for nineteen years before selling her to W. Sherwood of Blackwell, Co. Durham. Later she was sold to G.E. Turton of Kildale, in the North Riding of Yorkshire who last used her in 1944. Photographed at Sherwood's yard at Blackwell, Nr Darlington, on 25th May 1941.
*(author's collection)*

233. At Limerick Fair in 1924 is Fowler A9 compound engine No 15713■, *Pride of Shannon*, Registration No TI 1501, built in May 1924 and sold to David Donnelly of Limerick, Ireland. In 1934 she was sold to McMahon of Ballina, Co Mayo, staying in his ownership until whe was sold out of showland in 1965.
*(courtesy Leeds T.E. Club)*

234. Fowler B6 compound engine No 16103, *Lioness*, Registration No NY 5453, was built in January 1924 and sold to James Studt of Reading. Her next owner was Charles Studt of Pontypridd, Glamorgan who sold her to Charles Openshaw of Reading. By 1933 she was travelling in the ownership of William Murphy of Newcastle upon Tyne. Before the war she was sold to John Powell of Jarrow, Co Durham with whom she spent the rest of her working life. Photographed at Newcastle Town Moor Fair in June 1939. *(author's collection)*

235. Fowler B6 compound No 16103 *Lioness* at Darlington Carnival in 1933 whilst in William Murphy's ownership. *(author's collection)*

236. A late thirties photograph of Fowler B6 compound engine No 19782■, *The Lion*, Registration No UB 9763 at Babbacombe Fair, Devon. Built in March 1932 and sold to Anderton & Rowland of Bristol, this engine was the first Fowler for over 50 years to be fitted with left hand steering and was the last engine to be used regularly by the firm, being, in turn, with the scenic speedway and the ark. The latter was the last ride she attended before being laid up at Bristol, in 1950 she was sold out of showland.
*(courtesy J. Hobbs)*

237. Driver Dick Davies poses in front of his engine, Fowler No 19782■, *The Lion*, at Taunton Fair in 1945.
*(courtesy R. Davies)*

238. Fowler B6 compound 19989, *Onward*, Registration No XG 2010, built April 1933 and sold to Samuel N. Ingham of Hyde, Cheshire. In 1940 she was sold to Pickfords Ltd who she served for ten years before being cut up by Cohens in 1950. Of the four Super Lions built this was the only one that has not survived.
*(author's collection)*

239. Garrett No 4 compound tractor No 27185, Registration No BJ 623, was built in August 1908 and sold to John Studt of Cardiff. The following year she was returned to Garretts who then sold her to Mrs E. Smith of Devonport who named her *Teddy Bear*. By 1924 she was travelling under the banner of William Thurston of Cambridge and was used for pulling his son Stanley's living van. In 1932 whe was sold out of showland eventually being scrapped in 1942. Seen at a Devon Fair circa 1920 when owned by Mrs E. Smith & Sons.

*(courtesy D.E. Hattam)*

240. Garrett 7nhp compound engine No 27360, Registration No CK 3407, was built in March 1909 and sold to R.H. Hayne & Co of Ardmill. Her next owner was William Mitchell & Sons of Preston, Lancashire who converted her into a showmans engine. In 1921 she was bought by James Dewhurst & Sons of Preston, with whom she spent the rest of her working life until being laid aside in 1949. She had her uses even then as is shown in this photograph taken in James Dewhurst's yard at Preston in 1954. *(courtesy B.D. Stoyel)*

241. Garrett No 4 compound tractor No 28409, Registration No BJ 897 was built in June 1910 and sold to John Aird & Sons of Fulham, London. By 1913 she had been bought by Fletcher Bros of Cathays, Cardiff and converted into a showmans tractor which they named *Pride of Swansea*. Her second showland owner was William Davis of Bexhill on Sea, East Sussex. She was later sold out of showland. *(author's collection)*

242. Garrett No 4 compound tractor No 29771, Registration No BJ 1162, was built in September 1911 and sold to F. Maltravers & Son of Coggeshall, Essex. In 1917 she was sold to W. Johnson & Son of Braintree. In 1924 she was acquired by William Barker Junior of Stowmarket, East Suffolk who converted her into a showmans tractor and called her *Renown*.She was in his ownership when this photograph was taken at Christchurch Park Fair, Ipswich on 17th May 1934. In 1941 she had two more owners before being bought by Ernest H. Manning of Great Barford, Bedfordshire with whom she spent the rest of her working life. *(courtesy R.G. Pratt)*

243. Another photograph of Garrett No 29771, *Renown*, seen here with loads on the Woodbridge by-pass in 1935.
*(courtesy R.G. Pratt)*

244. Garrett No 4 compound tractor No 30159, Registration No BJ 1206, built in December 1911 and sold to Charles Purser of Cradley, Hereford. At a later date Jenning Bros of Devizes became her fourth owner, converted her into a showmans tractor and named her *Victoria*. She had two more owners before being scrapped at Southampton in the mid-fifties.
*(author's collection)*

245. Lying derelict at Tom Bugg's yard at Nayland on 13th March 1935 is Garrett No 4 compound tractor No 32868, Registration No BJ 3004. She was built in April 1916 and sold to W.S. Dewing of Norwich and in 1920 she was purchased by Thomas Saul & Sons Ltd of Great Yarmouth. In the course of the next few years she was sold to John Barker of Ipswich who converted her into a showmans tractor and then to Tom Bugg of Nayland, West Suffolk who travelled her until 1933.
*(courtesy R.G. Pratt)*

246. Garrett No 4 compound tractor No. 33305■, Registration No HT 7112, photographed at New Malden Fair in July 1934. She was built in June 1918 and sold to the Ministry of Munitions. By 1924 she was in the ownership of David Harrison of Bristol who converted her into a showmans tractor and travelled her with his chairoplanes. By 1925 she has been sold to James Edwards Junior of West Croydon, Surrey, who gave her the name *Mighty Atom*.
*(courtesy B.D. Stoyel)*

247. Liscard Park Fair on 6th July 1937 is the setting for this photograph of Garrett No 4 compound tractor No 33401, Registration No HR 101. Built in November 1918 she was sold to the Ministry of Munitions and a year later was in the ownership of Wort & Waybro of Salisbury. In the twenties she was acquired by Miles Jervis of Chasetown, Staffs, converted into a showmans tractor and given the name *Wait and See*. In the first part of 1930 she was sold to James P. Collins of Hightown, Manchester, later passing to his son John Collins, Junior. By 1938 she had been scrapped.
*(courtesy B.D. Stoyel)*

248. Driving the chairoplanes at Frome Fair is Garrett No 4 compound tractor No 33566■, Registration No BJ 4395. Built in July 1919, her first owners were Browning Bros of Stonehouse, Glos, who sold her to Workman Bros of Slimbridge before she was acquired by William Cole of Yate who converted her into a showmans tractor. She was named *Little Billie* and given a new Registration No AD 9069. For a number of years she travelled his chairoplanes and eventually she was taken over by Bernard Cole, William's son, and was used to travel his dodgems in company with Fowler 14880, *Princess Marina*. Her last owner was Percy Cole of Martock, Somerset. She was finally sold out of showland in the seventies.
*(courtesy J. Hobbs)*

249. McLaren 8nhp compound engine No 1053, *Magic*, Registration No NC 2027 was built in May 1909 as a road locomotive and sold to Norman E. Box of Ardwick, Manchester. Early in 1930 she was bought by Arthur Bates of Rhode Heath & Salford who converted her into a showmans engine to travel and power his Auto Scooter cars. At a later date she was sold out of showland to Sir Lindsay Parkinson & Co Ltd of Temple Newsham, Leeds. By 1948 she had been scrapped by Thomas Wards. Photographed at Nelson Fair on 15th July 1936 with Bates's ride. *(courtesy B.D. Stoyel)*

250. McLaren 5nhp compound tractor No 1301, Registration No ER 1331 was built in August 1911 and sold to Ironside, Son & Dykerhoff of London. Upon the outbreak of the First War she was acquired by the War Department. She is seen here on 26th June 1935 at Cambridge Midsummer Fair when she was in the ownership of James Harris of Gamlingay, Cambridgeshire (& Biggleswade, Bedfordshire). He bought the engine from the War Department in 1923 and had her converted into a showmans engine by Shepherd, Engineers of Luton. She was named *Lord Lascelles* and spent the rest of her working life with this concern. *(courtesy R.G. Pratt)*

255. Savage single cylinder traction engine *The Indian Chief* whilst in the ownership of Simons & Teago of Menai. A full length awning and dynamo have been fitted. No details are known about this engine as it is not listed in the existing Savage records. (courtesy B.D. Stoyel)

256. Savage 8nhp compound traction centre engine No 730, *Empress*, was built in 1898 and sold to Barker Thurston & Sons of Norwich. She is seen circa 1906 pausing at Olney in Buckinghamshire whilst hauling the roundabout loads. The roundabout was built up around the engine which provided the driving power. In place of a dynamo in front of the chimney was a small steam engine for driving the organ.

(author's collection)

257 & 258. Two interesting photographs of Savage traction centre engine No 730, *Empress*, in trouble at Wollaston, Northants, in 1906 when she ran off the road and needed the assistance of a local farmer to help pull her out. *(author's collection)*

259. Lying out of use at Sutton Coldfield in 1933 is Savage 7nhp traction centre engine No 847, *The Wonder*, Registration No DH 2478. She was built in January 1909 and sold to Pat Collins of Bloxwich in Staffordshire. This engine was fitted with duplex cylinders and carried neither dynamo nor organ engine. *(courtesy R. Smith)*

260. Tasker B2 compound tractor No 1595, Registration No AA 5264 was built in February 1914 and sold to Weymouth Steam Haulage Co. Her next owner was Hall & Co of Worthing before she was acquired by J. Matthews of West Worthing who converted her into a showmans tractor and named her *Pride of Sussex*. Her last owners were Robert & Gardiner of Stamford, Kesteven, who finished with her in 1953. *(author's collection)*

261. Tasker B2 compound tractor No 1596, Registration No AA 5262 was built in February 1914 and sold to F.W. Bloomfield of Wymering, Hants. Her next owner was Mrs J. Smith of Shirley, Southampton who converted her into a showmans tractor before selling her to J. Russett & Sons, also of Southampton, who named her *King of the Road*. She is seen here at Southampton Easter Fair in 1929. *(courtesy R. Smith)*

262. Tasker B2 compound tractor No 1770, Registration No HO 2285 was built in February 1919 and sold to Holloway Bros of Liverpool. She then went to Box & Turner of Ardingly, East Sussex before being bought by showman Tom Smith of Shoreham by Sea, West Sussex. He used her for haulage purposes only never fitting her with a dynamo, in fact his only concession to fairground decoration was the fitting of two twisted brasses to her front canopy supports. After he sold her she had two more owners and was last licenced in 1947. Seen at Lindfield Fair, Sussex on Saturday 12th August 1939 whilst in Tom Smith's ownership.
*(courtesy A.C. Durrant)*

263. Wallis & Steevens 7nhp compound engine No 2319, Registration No KE 3532 was built in April 1895 and sold to James Jennings Junior of Swindon, Wiltshire. Circa 1898 she went back to Wallis & Steevens who resold her in April 1898 to John Penfold of Plumstead Marsh, Kent: he named her *Sweet Duchess*. She spent the rest of her working life with this concern. She is seen on the road with his 'Electric Galloping Horses' ride. *(author's collection)*

264. Wallis & Steevens 8nhp compound engine No 2643, *Royal John*, Registration No HR 4127 was built in April 1903 and sold to Henry (known as Tom) Jennings of Devizes, Wiltshire for travelling his set of galloping horses roundabout. During the time she was with Jennings she had only one driver Sid Willis. In late 1920 she broke her crankshaft and was left in a field at Aggbourne, Berkshire. In 1922 she was sold to J. & H. Williams of Tadley and Newbury complete with the gallopers. She remained with the Williams family for the rest of her working days before being scrapped circa 1930. *(courtesy J. Hobbs)*

265. Wallis & Steevens 8nhp compound engine No 2643 *Royal John* outside the Crown Inn at Ludgershall, Wiltshire helping to pull out a tractor and trailer from the edge of the Crown pond which appears to be in flood. *(courtesy Leeds Traction Engine Club)*

266. Wallis & Steevens No 2643 *Royal John* on the road with Henry Jennings' set of galloping horses roundabout, circa 1909. Driver Sid Willis is seated on the rear wheel. *(courtesy J. Hobbs)*

267. Wallis & Steevens 7nhp compound engine No 7192 was built in September 1912 as a sprung traction engine and sold to S.J. Keast of Wadebridge, Cornwall. In 1915 she was sold to Mrs Amy Lock of Plymouth who converted her into a showmans engine and, in 1937, named her *Queen Elizabeth*. By 1942 she had been sold out of showland. *(courtesy J. Hobbs)*

268. Wallis & Steevens 5 ton wagon No 7404, *The Swift*, Registration No AA 5207, was built in October 1913 and sold to John May & Co Ltd of Basingstoke. In 1922 she was bought by J. & H. Williams of Tadley & Newbury and in her later days travelled with their chairoplanes. She is seen standing derelict in the yard of Lewis Brown of Headley, Hampshire, in April 1936. *(courtesy B.D. Stoyel)*

THE SWITCHBACK PAR EXCELLENC
QUEEN OF THE SOUTH